# *Experiments to Study Our Atmospheric Environment*

## Steven Businger

*Department of Meteorology*
*School of Ocean, Earth Science and Technology*
*University of Hawaii*

Prentice Hall
Upper Saddle River, New Jersey 07458

Library of Congress Cataloging-in-Publication Data

Businger, Steven.
    Experiments to study our atmospheric environment / Steven
Businger.
           p.   cm.
    ISBN 0-13-369232-9 (pbk.)
      1. Atmospheric physics—Observers' manuals.    2. Atmospheric
physics—Experiments.    I. Title.
QC871.4.B88    1996                                 95-40828
551.5' 078—dc20                                  CIP

Production Editor: *Naomi Sysak*
Managing Editor: *Kathleen Schiaparelli*
Director of Production: *David Riccardi*
Production Coordinator: *Trudy Pisciotti*
Acquisitions Editor: *Robert McConnin*
Cover Design: *Jayne Conte*

©1996 by Prentice-Hall, Inc.
Simon & Schuster Company/A Viacom Company
Upper Saddle River, New Jersey 07458

Printed in the United States of America

10  9  8  7  6  5  4  3  2  1

ISBN 0-13-369232-9

Prentice-Hall International (UK) Limited, *London*
Prentice-Hall of Australia Pty. Limited, *Sydney*
Prentice-Hall Canada Inc., *Toronto*
Prentice-Hall Hispanoamericana, S.A., *Mexico*
Prentice-Hall of India Private Limited, *New Delhi*
Prentice-Hall of Japan, Inc., *Tokyo*
Simon & Schuster Asia Pte. Ltd., *Singapore*
Editora Prentice-Hall do Brasil, Ltda., *Rio de Janeiro*

# PREFACE

This manual is designed to be used as a text in an introductory atmospheric sciences laboratory course. It takes a different approach from traditional workbooks used in many meteorology labs in that it comprises primarily hands-on experiments selected to convey fundamental concepts in meteorology. These experiments are designed give students of diverse academic backgrounds an opportunity to explore and understand the underlying physical principles of meteorology firsthand, and encourages them to apply these principles to better understand our everyday atmospheric environment. The materials for the experiments described can be obtained inexpensively from most scientific supply houses if not local stores. Some of the shorter experiments can be combined with others to take up the allotted laboratory period. Other activities, such as the analysis exercises in section 7, will take longer and a subset can be assigned.

A lesson consciously promoted by this manual is the art of making careful observations in the context of the scientific method: the cornerstone of modern science. Progress in science was revolutionized during the renaissance through development of the scientific method, in which careful observation leads to a prediction or hypothesis, followed by experimentation and finally the formulation of a theory. The following example of this combination of observation and reasoning applied in the scientific method is a classic experiment in atmospheric science: In 1643 Torricelli had invented the first barometer by immersing a glass tube in a dish of mercury and showing that a column of mercury would rise up the tube so as just to balance the weight of the air above the dish, thus providing the first measure of atmospheric pressure. Four years later, Pascal made one of the first completely scientific predictions. He reasoned that since atmospheric pressure is due to the weight of the air above you, there should be less pressure when you climb a mountain, and so the mercury in the barometer should not rise as high. Upon climbing the Puy du Dome in the French Alps with the help of friends, Pascal confirmed his prediction. News of his successful prediction caused great celebration all across Europe, because it represented a triumph of the new scientific method. By approaching the experiments in this manual with a spirit of inquiry like that of Pascal, the world of atmospheric science will open to the student.

## ACKNOWLEDGMENTS

The author is grateful to Mary McVicker and Diane Henderson for editing of the manuscript and Mary McVicker and Steven Chiswell for assistance with figure drafting. Thanks are also due to lab instructors at North Carolina State University and the University of Hawaii for field testing the exercises in this manual and offering suggestions for improvements. Finally, I am grateful to my wife Susan for her encouragement and patience during the creation of the manuscript.

## ABOUT THE AUTHOR

Steven Businger is an Associate Professor at the University of Hawaii. He has also served on the faculties of North Carolina State University and the University of Washington. Dr. Businger received his PhD in meteorology at the University of Washington where he studied the evolution of arctic storms. He has authored numerous articles on a range of subjects in the field of atmospheric science including, winter storms, severe thunderstorms, hurricanes, and acid rain. He has been active in teaching the fundamentals of meteorology since 1975.

# TABLE OF CONTENTS

# *EXPERIMENTS TO STUDY OUR ATMOSPHERIC ENVIRONMENT*

## **Steven Businger**

*Department of Meteorology*
*School of Ocean, Earth Science and Technology*
*University of Hawaii*

# Section 1  Origin and Composition of the Atmosphere

The Earth's present atmosphere is a mixture of gases, mostly nitrogen and oxygen, and large numbers of suspended particles. The concentrations of the most abundant components of the gaseous mixture, called air, are uniform through the lowest 80 km of the atmosphere and are essentially constant with time. On the other hand, concentrations of certain important gases such as water vapor and ozone vary substantially from place to place and from time to time. Carbon dioxide is fairly uniformly distributed, but its concentration has been increasing continually since before 1900. Although not a toxic gas, carbon dioxide interferes with the radiation balance of the Earth and may have an effect on the global climate.

The atmosphere contains huge numbers of particles from natural sources, such as the oceans and volcanoes, and from human inventions, such as motor vehicles, power stations, and industrial plants. A combination of air and the suspended particles is called an *aerosol*. Some particles, liquid and solid, play important roles in cloud, rain, and snow formation. Other particles

1

are hazardous to health and costly to society.

Compared to the radius of the Earth (6,370 km), the depth of the atmosphere is quite shallow.  About 99.92% of the mass of the atmosphere is below 50 km, and roughly half of the mass of the atmosphere is below about 5 km.

The weight of the air in a column of unit cross section and extending to the top of the atmosphere is known as *atmospheric pressure*. It is a maximum at the Earth's surface and always diminishes with increasing height because the mass of air above you decreases as you move to higher altitude.  At sea level, atmospheric pressure averages about 1,013 millibars (~29.9 inches of mercury). In English units, this amounts to ~14.7 pounds per square inch.  At an altitude of 50 km, the pressure is about 0.85 mb.  Contours, or lines of constant pressure shown on weather maps, are called *isobars*.

It is well-known that, on the average, air temperature decreases with height through the lower atmosphere.  This layer, where temperature decreases with height, is known as the *troposphere*.  It ends at a level called the tropopause and is surmounted by the *stratosphere*, through which the temperature is constant or increases with height.  Observations show several other higher layers of decreasing and increasing temperature (See Fig. 1.1).

In the stratosphere, chemical reactions, stimulated in part by the absorption of ultraviolet radiation from the sun, lead to the establishment and maintenance of a layer of ozone. This layer extends from about 10 to 50 km and is extremely important to life on Earth. The ozone greatly reduces the amount of ultraviolet radiation reaching the ground. Ultraviolet rays can cause skin cancer and affect other biological processes. Various substances, such as chloro-fluorocarbon gases, introduced into the atmosphere by human activities, appear to pose a threat to the ozone layer.

In the upper layers of the atmosphere, air densities and pressures are very low; gaseous atoms and molecules exist in relatively small concentrations. The absorption of ultraviolet radiation causes electrons to be stripped from some gaseous species. This process accounts for a deep region of charged particles known as the *ionosphere*. Some long-distance radio communications systems still depend on reflections from the ionosphere. On some occasions, the ionosphere is disturbed, and radio transmissions via the ionosphere become ineffective.

At the uppermost reaches of the atmosphere, high-speed particles from the sun are guided towards polar regions by the Earth's magnetic field. They collide with air molecules and cause electrons to be freed. When recombination occurs and the air molecules return to their original states, light is emitted. The resulting brilliant displays of light are called the *aurora borealis* in the Northern Hemisphere, the *aurora australis* in the Southern Hemisphere. The more common name for these two phenomena are the northern and southern lights, respectively.

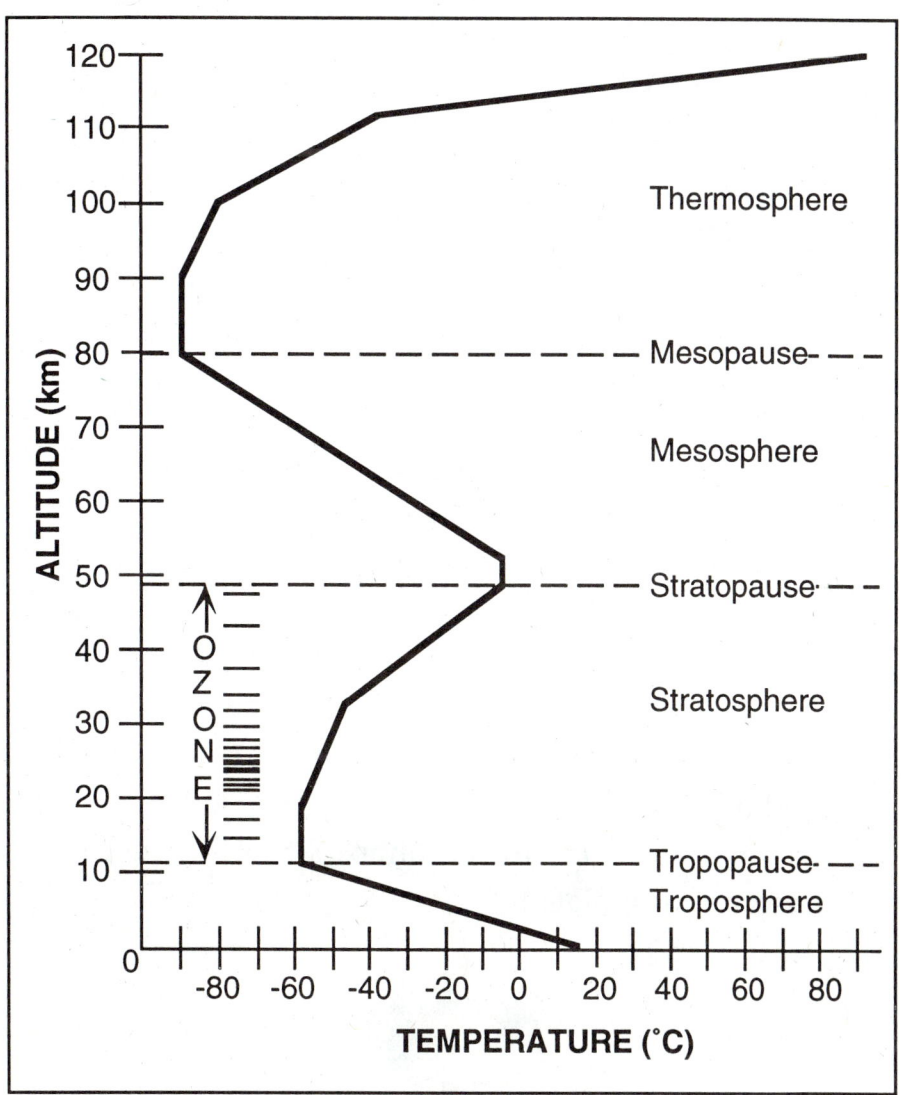

Figure 1.1 Temperature structure of the Earth's atmosphere

### *Origin of the Earth's Atmosphere*

The Earth's atmosphere originated from an outgassing of dissolved gases from inside the Earth as the planet cooled and the crust solidified. A clue to the composition of the early atmosphere can be obtained by analyzing the nature of the gases escaping from present volcanoes; these include nitrogen, $N_2$, water vapor, $H_2O$, sulfur dioxide, $SO_2$, carbon dioxide, $CO_2$, and various other trace gases such as argon. One gas that is very important in the Earth's present atmosphere but conspicuously absent in volcanic emissions is oxygen ($O_2$).

The question arises: Was the early Earth's atmosphere devoid of oxygen? Since oxygen is chemically active (it's an oxidizer that causes rust, for example), analysis of rocks that were exposed to the Earth's atmosphere very early in its evolution can give a clue as to the presence or absence of oxygen at that time. The fact that these ancient rocks show no evidence of oxidation suggests that the Earth's earliest atmosphere actually was devoid of oxygen.

How did oxygen get added to the Earth's atmosphere? Scientist have hypothesized that very intense solar radiation in the primitive atmosphere split the oxygen from the hydrogen in water. Another theory, which is supported by strong evidence, suggests that the main source of oxygen in the Earth's atmosphere came from plant life. In the process called photosynthesis, plant life consumes carbon dioxide and emits oxygen. Once oxygen appeared, ozone ($O_3$) also appeared. In the geologic record there is evidence for a very rapid expansion of the plant life over the Earth's surface once oxygen and ozone appeared in the atmosphere. This suggests that the addition of oxygen and ozone to the atmosphere occurred very quickly following the emergence of plant life in the oceans.

# Lab 1:  The Percentage of Oxygen in the Atmosphere

## INTRODUCTION

Earth's atmosphere is composed of a mixture of gases, but ~99% of the atmosphere is made up of just two gases- oxygen and nitrogen.  Because many people have heard so much about oxygen and because they know they must breathe it in order to live, it is often thought that the atmosphere is all or mostly oxygen.  Another way that the presence of oxygen is evident is the oxidation process.  Oxygen has the ability to rust metals and make apples turn brown.  The oxidation process, in fact, changes a substance and often creates a new one altogether.  Although oxygen seems to be predominant in the atmosphere, the actual amount may surprise you.

## ACTIVITY

OBJECTIVE: The objective of this experiment is to determine the amount of oxygen in the atmosphere.

MATERIALS:
§ birthday candle
§ clay
§ petri dish (or small tin can)
§ 100-ml graduated cylinder
§ matches
§ glass marking pencils
§ large test tube
§ penny
§ water
§ safety goggles
§ food coloring

## PROCEDURE:

1. Place a small amount of clay onto the penny. Make sure that the clay does not extend over the edge of the penny and secure the candle in the clay as shown in Fig. 1.2. (If clay is not available, wax from a lit birthday candle can be dripped onto the penny, and the candle secured in the wax as it cools).

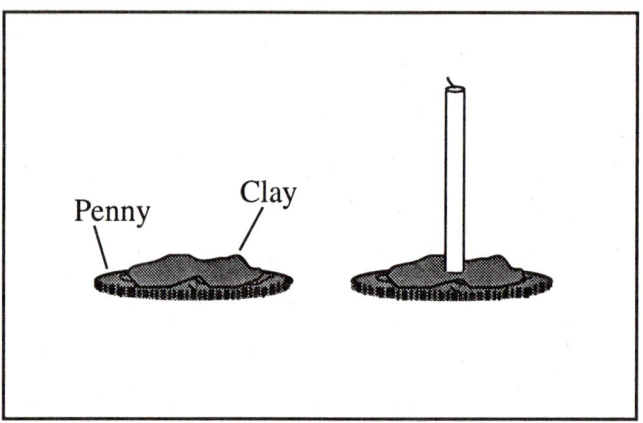

Figure 1.2

2. Fill the test tube with water. Using the graduated cylinder, measure this volume of water and record it in the Data Table. This volume is also the volume of air in the test tube when the test tube is empty.

3. Pour the water into the petri dish and add a drop of food coloring. Carefully place the penny and the candle in the center of the petri dish as shown in Fig. 1.3. (A small tin can may be used as a substitute for the petri dish.) The penny should keep the candle upright. In the space below predict what will happen if you light the candle and then place the test tube over the candle. **Explain your prediction.**

4. Holding the test tube in one hand, light the candle and then invert the test tube over the candle as shown in Fig. 1.4. Do this quickly, without touching the test tube to the candle. Make sure that the mouth of the test tube is below the surface of the water, but above the bottom of the petri dish.

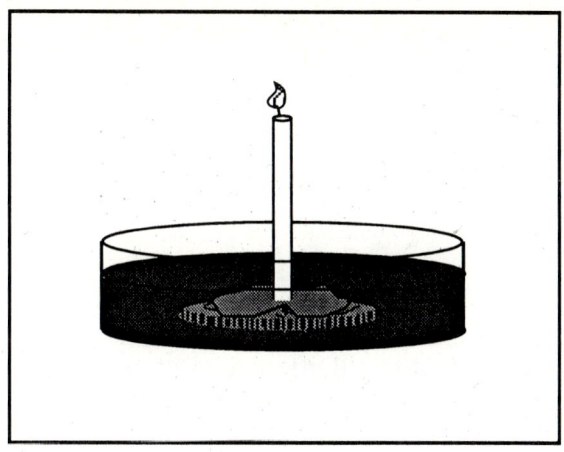

Figure 1.3

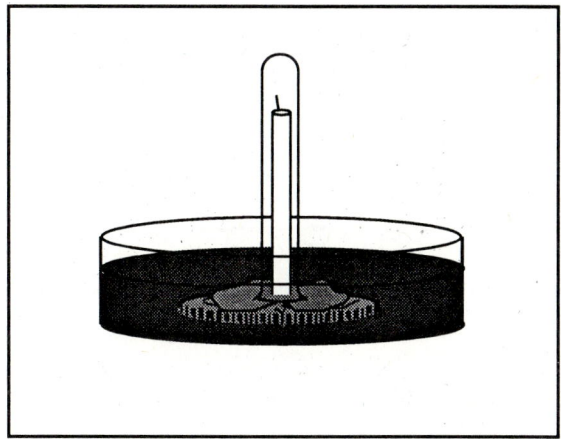

Figure 1.4

5. After the candle goes out, mark the level of the water in the test tube with the glass marking pencil. Then remove the test tube.
6. Fill the test tube with water up to the mark you just made. Measure the volume with the graduated cylinder and record it in the Data Table.
7. Subtract the second volume you measured from the first and record it in the Data Table.
8. To calculate the percentage of oxygen in the air, use this equation:

percentage of $O_2$ in air $= \dfrac{(\text{1st volume} - \text{2nd volume}) \times 100}{\text{volume of the test tube at start}}$

9. Wipe the mark off the test tube and repeat steps 4 through 8 two more times recording all data in the Data Table.
10. From your three trials, calculate an average percentage of oxygen in the air.

## Data Table

| Trial | 1st Volume of Test Tube | 2nd Volume of Test Tube | Difference Between 1st & 2nd Volumes | % of $O_2$ in Air | Average % of $O_2$ in Air |
|-------|-------------------------|-------------------------|--------------------------------------|-------------------|---------------------------|
| 1     |                         |                         |                                      |                   |                           |
| 2     |                         |                         |                                      |                   |                           |
| 3     |                         |                         |                                      |                   |                           |
|       |                         |                         |                                      |                   |                           |

QUESTIONS:
1. Why did the candle go out after the test tube was placed over it?

2. Why did the water rise up into the test tube when the test tube was placed over the candle?

3. Explain why the equation in Step 8 above can be used to calculate the percentage of oxygen in air.

4. Why is it important to do more than one trial?

5. How does the percentage of oxygen in the air compare with what you thought it would be? How can you explain the difference?

6. The actual percentage of oxygen in the atmosphere is 20.9%. If the percentage you calculated is different, what things might account for the difference?

7. What do you think would have happened if a larger test tube had been used? Explain your answer.

8. Refer back to your answer to question 1. Why do you think it is important to close the windows if there is a fire drill?

Part II: Toss a piece of burning notebook paper into a pirex flask or jar with a mouth slightly smaller than an egg. Place the peeled egg on top of the flask's mouth and carefully watch the egg when it is first placed on the flask (Fig. 1.5).

Fig. 1.5

1. When does the egg enter the flask?

2. Why does the egg enter the flask?

3. What will happen if the flask is inverted and heated gently with a candle? Why?

# *Lab 2: Particles in the Air*

## INTRODUCTION

The atmosphere contains great numbers of suspended particles from natural sources, such as the oceans (salt), forests (pollen) and volcanoes (ash), from human inventions, such as motor vehicles, power stations and industrial plants, and from human activities such as agriculture. A combination of air and the suspended particles is called an *aerosol*. Some particles, liquid and solid, play important roles in cloud, rain, and snow formation. Other particles are hazardous to health and costly to society.

Since suspended particles are difficult to see it is surprising how many there actually are in the air. In polluted air over land there may be more than 1000 particles per cubic cm, while in clean air over the ocean the count may be as few as 100 per cubic cm.

## ACTIVITY

OBJECTIVE: The objective of this activity is to investigate the amount and types of particulate matter in the air.

MATERIALS:
§ coffee filter
§ strainer
§ magnifying glass
§ 9 inch pie plate
§ white paper plate
§ filtered or distilled water

PROCEDURE:
**Note:** Some of this activity will be done at home.

1. Fill the pie plate full of water, and set it outside in a place that is exposed to the open air, and where the plate will not be disturbed. Leave the plate in place for 48 hours. If there is precipitation forecast in your area during the period of your experiment you can eliminate the distilled water and just filter the collected rainwater or melted snow water.

2. At the end of this time, take the coffee filter and put it in the strainer. If you have an automatic drip coffee maker, the filter holder will work best for this.
3. Take the filter and the strainer outside to the plate and carefully pour the water through the filter.
4. Rinse the sides of the plate with a small amount of water and pour this through the filter.
5. Carefully remove the filter and spread it out on the plate.
6. Allow the plate to sit undisturbed until the filter is completely dry. Be sure that there is no breeze near the filter while it dries. If there is a breeze, it will blow the particles away.
7. After the filter has dried completely, carefully fold it up and put it in an envelope. This will prevent the loss of particles while transporting it to class.
8. In class, use the magnifying glass to examine the particles on the filter. Without the magnifying glass, you may miss a lot of the particles.

QUESTIONS:

1. Describe what you see on the filter. What was your reaction when you made observations of the coffee filter? Were there more particles than you expected? Fewer? What about the types of particles? The colors, the sizes?

2. Compare your filter with your classmate's filter. *Explain some reasons why the particles on the filters may appear different.*

3. If there was precipitation in your area how might the results differ from those during a dry period?

4. Where do you think the different particles came from?

5. Some of the beneficial aspects of particles in the air have already been described. What harmful effects of these particles can you imagine?

6. What time of the year do you think you would have found more particles on the filter? Fewer? Explain

5. Are there areas of your city or town where you think particulate air pollution would be worse than where you did the activity? If so, where, and why do you think this?

6. Extra Credit: What weather conditions would increase the amount of particles in the atmosphere? Decrease the amount?

## *Section 2  Pressure And Temperature*

Pressure and temperature are two of the fundamental variables used to describe the state of the atmosphere.  Their vertical distribution in the atmosphere was mentioned briefly in Section 1.  The relationship between pressure and temperature is important in understanding atmospheric motions, storm systems, and climate.  Therefore a short summary is presented here.

### *Pressure*

As stated in Section 1, atmospheric pressure is the weight exerted by the overlying air molecules in a column of unit cross section extending upward to the top of the atmosphere.  Pressure can also be described as the force per unit area exerted by the continuous collisions of gas molecules.  How are these two seemingly disparate descriptions reconciled?  The first one looks at pressure from a larger perspective: weight is a force resulting from the acceleration due to gravity acting on the mass of the overlying air.  The latter description is from a molecular perspective: air pressure is proportional to the *speed and mass of the air molecules*, and the *frequency of their impacts*. Since the force due to atmospheric pressure is the same on every part of your body, you don't feel it as weight. This fact led the ancients to hypothesize that air had no weight at all.

The weight of the atmosphere can be measured by means of a *barometer* (first invented by Torricelli in 1643). Liquid mercury is carefully poured into a "J"-shaped glass tube and then the tube is inverted, creating a vacuum space at the top of the tube. The height to which the mercury rises in the tube due to the weight of the atmosphere is a measure of the atmospheric pressure. If one were to carry a barometer up the side of a mountain, the height of the mercury column would gradually decrease, reflecting the fact that there is less air above the level of the barometer as its elevation increases.

### Temperature

Temperature can be thought of as the degree of hotness or coldness of an object. *Temperature is proportional to the speed and mass of the air molecules* (their average kinetic energy). From the molecular view it is clear that there must be a close relationship between pressure and temperature, since pressure is also related to the speed and mass of air molecules, as well as the frequency of their impacts. A simple experiment can be conducted to clarify this relationship. Take an air-tight container (constant volume of air) and insert a thermometer and a pressure gauge. Then place the container on a source of heat (e.g., burner). We find that as the temperature rises the pressure also rises. When a graph is made of the pressure versus the temperature the result is a straight line (a linear relationship between pressure and temperature).

When this straight line on the graph is extended to the pressure-axis where pressure equals zero, the coldest temperature theoretically possible is reached. This temperature is referred to as zero *Kelvin* (in honor of the scientist who made this discovery). Kelvin units are abbreviated K. A temperature scale starting with zero K, thus eliminating negative temperatures, is used by scientists for all calculations relating temperature and pressure. Thus, in calculating pressure from temperature using Kelvin, the result will always be greater than zero, consistent with the observation that a negative pressure value is without meaning.

Two other commonly used scales are the Fahrenheit scale and the Celsius scale. The Fahrenheit scale was named for G. Daniel Fahrenheit, the scientist who developed it. The value of 32 was assigned to the temperature at which water freezes and the value of 212 was assigned to the temperature at which water boils. The value of 0 was the lowest value Fahrenheit obtained in his experiments with a combination of salt water and straw. Fahrenheit used the human body temperature of near 100°F (98.6°F) to fix a second point on his scale. Between the freezing point and the boiling point are 180 equal divisions, each called a degree.

The Celsius scale was developed after the Fahrenheit scale, but in a similar fashion. A Swedish astronomer, Anders Celsius, assigned the values of 0 and 100 to the freezing and boiling points of water. The distance between the two points was divided into 100 equal increments, also called degrees. Each degree Celsius is 1.8 times larger than a degree Fahrenheit.

The thermometer as a meteorological instrument has a long history. An early version was first constructed by a Greek scientist named Hero about 2000 years ago. A water thermometer was invented in 1597 by Galileo, which must have broken with every freeze. Later wine and mercury were used as the expanding liquid in the thermometer.

Temperature variations across the Earth's surface are also associated with differential heating (land, sea), altitude, latitude (geographic position), and ocean currents. Contours of constant temperature on weather maps are referred to as isotherms. The subject of winds will be discussed further in Section 7.

### Heat

*Heat is that which makes things hotter*, the transfer of energy from one object to another. Heat is measured in units called *calories*. A calorie is the amount of heat (or energy) that is required to raise one cubic centimeter of water by one degree centigrade. For a given substance, the temperature change caused by any quantity of heat depends on the mass of material involved and the *specific heat*. The specific heat is the quantity of heat that must be added to a unit mass of a substance in order to raise its temperature 1°C. Thus, the specific heat of water is 1 cal/gm °C (or the magnitude of the specific heat of water is 1).

There are four ways in which heat is commonly transferred in the atmosphere.  *Conduction* is the transfer of heat between molecules within a substance.  When you leave a metal spoon in a pot of boiling water, the spoon becomes hotter.  This is an example of conduction.  Wind chill is an example of conduction in the atmosphere.

Another form of heat transfer is *convection*.  The transfer of heat by the mass movement of a fluid is called convection.  Convection occurs mostly in liquids and gases.  When a pot of water is boiling, convection is occurring.  In the atmosphere, thunderstorms are a vigorous form of convection that transfers heat and moisture from the surface into the atmosphere.

A third form of heat transfer is *radiation*.  A large amount of heat is received from the sun in the form of radiation.  Energy travels from the sun to the Earth in the form of electromagnetic waves, which are largely absorbed at the Earth's surface.  Heat from a fire is another example of energy transfer by radiation.  An expanded discussion of radiation is given in Section 4.

The last method in which heat is transferred in the atmosphere is often overlooked.  It is the *latent heat* associated with changes in phase of water.  For example, heat energy is required to evaporate water.  The energy in the form of latent heat is released in convective updrafts when water vapor condenses into cloud droplets.  This added heat increases the buoyancy of clouds, which acts as a driving force for thunderstorms.  An expanded discussion of latent heat is given in Section 4.

# Lab 3: Atmospheric Pressure

## INTRODUCTION

Contrary to our intuition, air pressure is exerted equally in all directions: down, up, and sideways. Therefore, when you are inside a building, air pressure is not just the weight of all of the air within the room up to the ceiling, but it is the weight of the air in a column above the building that extends to the top of the atmosphere.

## ACTIVITY

OBJECTIVE: The purpose of this activity is to demonstrate the nature of atmospheric pressure.

MATERIALS:
§ stiff plastic cup or a glass
§ stiff paper cup
§ cardboard
§ straight pin
§ water

PROCEDURE:
1. In doing this activity, work over a sink or catch basin. Fill the plastic cup (or glass) up to the rim with water. Cover the cup with the cardboard. *In the space below, predict what will happen if you turn the cup over with the cardboard covering the cup, and explain your prediction.*
2. While holding the cardboard onto the cup, carefully turn the cup over. Hold the cup by the bottom and release the cardboard as shown in Fig. 2.1. In doing this part of the activity it is important that the cup is not deformed in the process of turning it over. Why is this so?
3. Slowly, turn the cup sideways. *What happens?*
5. Use the straight pin and carefully make a hole in the bottom of the cup. Then carefully repeat the process while holding a finger over the hole in the bottom of the cup. Lift your finger. *Explain the result.*

Repeat the same process with the paper cup. *Were the results any different?*

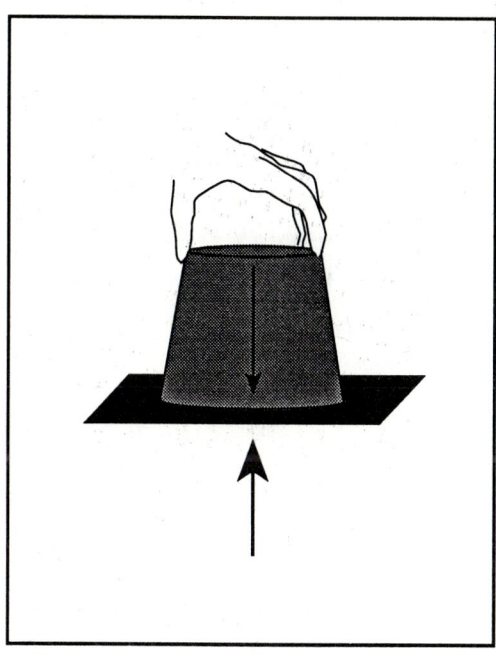

Fig. 2.1

QUESTIONS:

1. In this activity, what held the cardboard to the cup, preventing the water from falling out of the cup?

2. Explain why the water and the cardboard fell from the cup when the straight pin was inserted into the paper cup.

# *Lab 4: Part I: Crushing Cans*

## INTRODUCTION

This activity provides a dramatic demonstration of the force that is exerted by atmospheric pressure. This experiment works well as a demonstration. If done correctly, it will produce dramatic effects. If, however, the experiment fails, the activity is easily and quickly repeated. (This activity can also be performed using a larger gasoline can - without the gasoline naturally.)

## ACTIVITY

### OBJECTIVE:

The purpose of this lab is to explore the relationship between pressure and temperature, and the magnitude of the force represented by atmospheric pressure.

### MATERIALS:

§ aluminum can
§ pan
§ tongs
§ water
§ gas burner

### PROCEDURE:

1. Put cold water and ice in the bottom of the pan as shown in Fig. 2.2.
2. Put a tablespoon or so of water in the bottom of the aluminum can and heat the can until a cloud appears. The can needs to be rather hot for this experiment to work well (use the presence of steam as a guide). **CAUTION: *do not*** touch the can with your bare hands.
3. When wisps of cloud* appear at the opening of the can, use the tongs and quickly invert the can in the pan of water. (*Sometimes people mistakenly refer to this cloud as steam. Actually steam is invisible. It is not until the vapor cools and condensation takes place that a cloud becomes visible. See section 6 for more on this subject)
4. Repeat steps 1 through 3, using water at room temperature in the pan. Can

you predict the outcome of the modified experiment?

Figure 2.2

QUESTIONS:

1. What happened to the can in this activity? Explain.

2. When the can emerges from the water, it is practically filled with water. Explain why.

3. How did the result differ when water at room temperature was used in the pan? Explain.

4. The tablespoon of water placed in the can in step 1 plays an important role in this experiment. Explain.

# *Lab 4: Part II: Exploding Popcorn*

## INTRODUCTION

Dramatic pressure changes occur when water changes phase from liquid to vapor. This is one of the reasons why water boils at 100 °C at sea level. At this temperature the vapor pressure of water is the same as the atmospheric presssure at sea level. As more heat is added to the pot of boiling water, the temperature of the water stays at 100 °C. The added energy goes into breaking the bonds between the water molecules as they are released as individual vapor molecules. However, if the pot has a tight lid (as in the case of a pressure cooker), the temperature of the water can continue to rise, and consequently the food will cook faster. Eventually, the pot would explode if enough energy was added and it did not have a safety valve to release the pressure. This is what happens in the case of some volcanic eruptions, such as Mt. St. Helens (see Lab 38). Water, from snowmelt, percolates into the neck of the volcano where it is heated by the hot magma within. The overlying rock acts as a lid, which finally is displaced by a small earthquate or tremor, releasing a tremendous built-up of pressure in a catestrophic explosion.

A single kernel of popcorn can provide some interesting pressure physics to demonstrate these principles. A popcorn kernel is made up of starch, protein, fat, minerals, and water. The presence of liquid water is critical to popping. When popcorn is heated, the water inside the kernel becomes vapor and the internal pressure increases to as much as 9 atmospheres (1 atmosphere ~ mean sea level pressure). When the hull can no longer withstand the pressure difference between the pressure inside and the air pressure outside, the kernel explodes or pops.

## ACTIVITY

OBJECTIVE: To determine the moisture content of popcorn kernels and to observe how moisture and pressure relate to the popcorn's ability to pop.

EQUIPMENT/MATERIALS:
§ normal popcorn
§ oven-heated popcorn

§ 125-ml Erlenmeyer flask
§ aluminum foil
§ Bunsen burner
§ ring stand and ring
§ wire gauze
§ ruler
§ permanent felt marking pen
§ balance
§ 10-ml graduated cylinder
§ heat-resistant glove or flask tongs
§ 100-ml graduated cylinder

PROCEDURE:

1. Using aluminum foil, fashion a lid for a 125-ml flask. Determine the mass of the empty flask and lid. Record it in the following Data Table. Be sure to put units on the data.

2. Using the 10-ml graduated cylinder, measure the volume occupied by 20 kernels of unpopped normal popcorn. Record your findings in your Data Table.

3. Put the 20 kernels of unpopped corn in the Erlenmeyer flask and determine the mass of the flask, popcorn and the foil lid. Record this information.

4. Using subtraction, determine the mass of the 20 kernels of unpopped corn. Record this data.

5. Make sure the foil lid fits snugly over the mouth of the flask. With a pencil or a pen, make a few small holes in the foil.

6. Heat the flask over the Bunsen burner to pop the corn. Be sure to use wire gauze between the flask and the burner. Let the flask sit undisturbed until the kernels start to change color. Then use flask tongs or heat-resistant gloves to hold the flask. Shake the flask over the burner until all kernels pop.

7. Remove the flask from the heat and carefully remove the foil. CAUTION! Do not get burned by the steam! Let the flask stand for a few minutes to cool.

8. When the system is cool, determine the mass of the flask, lid and popped corn. Record this information.

9. Mark a kernel of oven-heated popcorn with a felt marker and place it in a 125-ml Erlenmeyer flask along with a normal kernel. Cover the flask tightly with a foil lid. Heat the flask as instructed before. Watch the two kernels to determine which pops first and which produces the fluffiest popcorn. Record these observations.

Measure in millimeters the longest side of each of the two popped kernels. Record this information.

## Data Table

| | |
|---|---|
| MASS OF EMPTY FLASK AND LID | |
| VOLUME OF 20 UNPOPPED NORMAL POPCORN | |
| MASS OF FLASK, LID, UNPOPPED POPCORN | |
| MASS OF UNPOPPED POPCORN | |
| MASS OF FLASK, LID, POPPED CORN | |
| FIRST TO POP: HEATED OR NORMAL? | |
| FLUFFIEST: HEATED OR NORMAL? | |
| LONGEST SIDE OF HEATED (mm) | |
| LONGEST SIDE OF NORMAL (mm) | |

QUESTIONS:

1. Why is the mass in popped corn less than the mass in the unpopped corn (i.e., what has left the system)?

2. What is the mass of the water lost by the system? Show mathematically how you arrived at this answer.

3. What was the percentage of water in the unpopped kernels? Show mathematically how you arrived at this answer.

4. The ratio of the volume of the popped corn to the volume of the unpopped corn is called expansion volume. The popcorn industry uses expansion volume as a test of quality. Orville Redenbacher claims his gourmet popping corn has an expansion volume of 40 to 1. What is the expansion volume (expansion ratio) for the popcorn used in this experiment? Show mathematically how you arrived at this answer.

5. What substance is present in a larger amount in the normal kernel as compared with the oven-heated kernel? In which kernel was the internal pressure greater before popping?

6. Which kernel produces more desirable popcorn, the normal or the oven-heated?

7. Reread the lab introduction and answer the following questions: Would popcorn pop faster on Pike's Peak or in Boston? Why?

8. Prepare a graph of expansion volume (dependent variable with no units) as a function of the percentage of water (independent variable with % units). A Data Table must accompany the graph. Use: Expansion Volume, Percentage Water (%) and Source of Data, as column titles. The Source of Data column should identify the lab partnership from whom the data was obtained.

Use data from the whole class and make the graph cover as much of a piece of graph paper as possible. Draw a best-fit line on your graph. *Remember:* Label the axes of the graph and show the units where appropriate. Protect the points on your graph with circles, squares or triangles. Title and number your graph.

# Lab 5:  Density, Temperature, and Pressure

## INTRODUCTION

In Lab 4, a closed container (constant volume) was used.  In the open atmosphere the relationship between air pressure, temperature, and volume is somewhat more complex.  When the temperature of the air changes, so does its density.  Molecules move closer together or farther apart as the temperature increases or decreases.  Therefore, when air is heated it expands, becoming less dense than the surrounding air and rising relative to the surrounding air.  Conversely, when air is cooled it contracts, becoming more dense than surrounding air and therefore sinks.

Within the atmosphere, when air temperature is increased, the expanding air mass rises, creating a region of low surface pressure.  When air temperature is decreased, the contracting air mass sinks, creating a region of high surface air pressure (recall that surface pressure is just the total weight of all the overlying air molecules. Since colder air molecules take up less space than warm ones, the total number of overlying air molecules in a cold air mass is larger and the surface pressure is higher).  This relationship between air temperature and the air pressure in an air mass is counter to the result of the closed container experiment, and thus a source of confusion.

## ACTIVITY

OBJECTIVE: The objective of this activity is to investigate the relationship of temperature and density.

MATERIALS:
§ balloon
§ empty 750-ml bottle
§ bucket of ice
§ bucket of hot water

## PROCEDURE:

1. Place the uncovered bottle in the bucket of hot water for three minutes. Do not submerge the bottle or allow water to get into the bottle.
2. Place the balloon over the mouth of the bottle. You now have an isolated mass of air. It is important to remember throughout this experiment that the amount, or mass, of air will remain constant. In the space below, predict what will happen to the balloon if the bottle is placed in a bucket of ice water. Explain your prediction.
3. Place the bottle in a bucket of ice water for three minutes. What happens? Remembering that the mass of the air has remained constant, what has changed?
4. Take the balloon off the bottle and place the bottle back in the bucket of ice for three minutes. Do not submerge the bottle or allow water to get into the bottle.
5. Place the balloon over the mouth of the bottle. As before, you have an isolated air mass. In the space below, predict what will happen to the balloon if the balloon is placed in a bucket of hot water. **Explain your prediction.**
6. Place the bottle in a bucket of hot water for three minutes. What happens? Remembering that the mass of the air has remained constant, what has changed?

## QUESTIONS:

1. What are the two variables of air that are being changed in this activity? What is the relationship between these two variables?

2. As the air is heated, what affect does this have on the air molecules?

3. In the atmosphere, what would you expect to happen to air that is warmed? Cooled?

4. Based on your observations and your answers to these questions, do you think it would be best to place a warm air vent near the floor or ceiling of a room? Explain your answer.

5. Extra Credit: As a balloon rises it expands and the air inside it cools. Explain why these observations are not contradictory.

The Earth moves around the sun in a nearly circular orbit with a period of about 365 days. At the same time, the Earth revolves about its axis once every 24 hours. The Earth's equitorial plane is tilted at an angle of 23.5° with respect to the plane of the orbit. Because of this tilt, the Northern Hemisphere intercepts maximum quantities of solar radiation during the middle months of the year, May, June and July, and the Southern Hemisphere experiences maximum solar radiation during the months of November, December, and January.  The seasons of the year are clearly related to these observations.

In explaining the seasons, it is common to consider the position of the sun with respect to the Earth, rather than vice-versa. On ~ June 21, the sun is at its most northerly position, located over latitude 23.5°N at noon. This date is called the *summer solstice* in the Northern Hemisphere. Six months later, the *winter solstice* occurs as the sun is over 23.5°S latitude at noon. The sun moves southward from June to December and is found over the equator on ~ September 22, the *autumnal equinox.* This term comes from the fact that, on that date, the length of day and night is equal at all latitudes. The same is true on ~ March 21, the time of *vernal equinox*, when the sun passes over the equator in its northward passage. The length of the day, the length of the path that sunlight must traverse through the atmosphere, and the surface area over which the radiation is distributed all vary with latitude and season and contribute to the balance of radiations

Radiant energy from the sun is in the form of *electromagnetic waves* and travels at the speed of light, $3 \times 10^8$ m/sec. The wavelengths of the sun's radiation cover a wide spectrum. Of particular importance in the atmospheric sciences are the ultraviolet band (wavelengths 0.01 to 0.4 µm), the visible band (0.4 to 0.7 µm) and the infrared band (0.8 to 1000 µm). Most solar energy is found in the visible and infrared bands.

The characteristics of the energy radiated by any body can be stated in three basic laws, which are derived from the more general Planck's law:

(1) *Stefan-Boltzmann law* - All objects emit radiant energy (except at 0 K), with hotter objects emitting more energy per unit area than colder objects.

(2) *Kirchoff's law* - Objects that are good absorbers of radiation are good emitters of radiation.

(3) *Wien's law* - The hotter the object the shorter the wavelength of the maximum emitted radiation.

The Stefan-Boltzmann law shows that the quantity of energy emitted varies with the fourth power of the absolute temperature, $E = cT^4$, where c is a constant. Therefore, if one body has a temperature of 6000 K (the sun) and another body has a temperature of 300 K (the Earth), the first one radiates $(6000/300)^4 = 160,000$ times more energy than the second for each unit area of surface.

A body which radiates energy according to the Stefan-Boltzmann law is called a blackbody. Most substances do not radiate as much energy as a blackbody does, since this represents a theoretical upper limit. The ratio of actual emission to blackbody emission is called the emissivity. It depends on the nature of the substance and may vary with wavelength and, to a lesser extent, with temperature. For example, snow has a high emissivity at infrared wavelengths, but a low emissivity at visible wavelengths.

A radiation law known as Kirchoff's law states that for any wavelength a substance's emissivity equals its absorptivity. The ratio of the actual energy absorbed to the total amount of energy intercepted by the substance is known as absorptivity. Kirchoff's law indicates that the absorptivity of snow is high at infrared wavelengths, but low at visible wavelengths. This accounts for the fact that snow reflects sunlight effectively and looks white.

The wavelength at which the emission of radiant energy is a maximum ($\lambda_{max}$) can be calculated from Wien's law. Specifically, the law is $\lambda_{max} = 2,880/T$, where T is the absolute temperature and $l_{max}$. is in micrometers. Taking the sun's temperature at 6000 K, Wien's law indicates that maximum solar radiation is at a wavelength of 0.48 μm, which is in the visible band.

The quantity of radiant energy incident on the top of the Earth's atmosphere is called the *solar constant* and amounts to ~1400 watts per square meter (~2 calories per $cm^2$ per minute). The absorption, reflection, and scattering of solar radiation as it passes through the atmosphere depend on the wavelength. Most of the ultraviolet radiation is absorbed in the upper atmosphere. It leads to photo-ionization and to ozone formation and destruction. Some visible radiation is reflected from cloud tops and scattered by air molecules and atmospheric particles. The fraction of incident visible solar radiation which is reflected back to space is called the Earth's *albedo*. The albedo is difficult to measure accurately but is considered to average ~30 percent. Most of the visible radiation passes through the atmosphere and warms the Earth's surface.

The *greenhouse effect* is the term used to describe the role of gases such as water vapor, carbon dioxide, and methane in increasing the temperature of the Earth's surface. These gases permit solar energy to pass rather freely through the atmosphere to the ground. But the infrared radiation emitted by the Earth is partly absorbed by these gases and prevented from escaping directly to outer space. It is important to note that water vapor is the principal greenhouse gas.

An examination of the radiation budget of the Earth shows that at the surface the amount of radiation absorbed exceeds the amount emitted at tropical latitudes. The net radiation, or radiation balance as it is called, is positive. Conversely, the radiation balance at polar latitudes is negative; more radiation is lost to space than is available from sunlight.

The major mechanisms for heat transport from low to high latitudes are air and ocean currents. Much of the transfer occurs as warm air and water move poleward while cold fluids move southward. A substantial quantity of heat is transported as latent heat. As water is evaporated at low latitudes, heat is added to the atmosphere in latent form. It is released as heat in storm systems when condensation leads to clouds and precipitation. Latent heat is explored in detail in Section 4.

Air temperatures at the Earth's surface depend on various factors, including the radiation balance. In any particular region, the albedo of the surface is an important variable. A land area composed of highly absorbing material will have a lower albedo than a sandy or snow-covered area. The albedo of water depends on the altitude of the sun. The average albedo of the ocean is about 8%.

The specific heat of water is much greater than that of land material, such as rock and sand. Thus, for the same mass, land regions undergo greater temperature changes than water regions when equal amounts of heat are added or taken away.

Radiant energy falling on land is absorbed in the top few millimeters and is slowly conducted downwards. Radiation incident on a water surface penetrates to substantial depths. Water currents further distribute heat in water bodies and evaporation consumes heat energy at the surface. These various factors largely account for the fact that continents are colder than the oceans in winter and warmer in summer. In fact, ocean temperatures change slowly and by small amounts throughout the year. The oceans store tremendous quantities of heat and act as a thermostat for reducing global temperature changes.

# *Lab 6: Distance from the Sun*

## INTRODUCTION

The fact that Earth is the only planet in the solar system that supports life is a direct consequence of its distance from the sun. If Earth were only two percent of its present distance farther away from the sun, it would be like Mars, a permanent "Ice Age" wasteland with a carbon dioxide atmosphere and all of its water tied up in polar ice caps. If Earth were only five percent closer to the sun, it would be like Venus, a planet many astronomers have described as a "hellish place". The surface temperature on Venus is about 850°F. Earth's distance from the sun is just right, and practically no other distance will do. Only recently, it has been determined that the range of distances from the sun in which Earth's conditions could have formed is very small compared to the scale of the solar system. Because of this narrow range (or "clement zone"), Earth's atmosphere is the only one in the solar system which will allow water to exist in all three states simultaneously- solid, liquid and gas.

This activity is designed to show how distance from a light source will affect temperature and that the range of distances in which a specific temperature can exist is relatively small.

## ACTIVITY

OBJECTIVE: The objective of this activity is to investigate the relationship between distance from a light source and temperature.

MATERIALS:
§ ruler
§ 4 identical thermometers
§ reflector lamp
§ clay or tape
§ watch

## PROCEDURE:

1. Place the ruler on a table and attach each thermometer to it with clay or tape at the correct distance (Table 3.1) to represent each planet (Fig. 3.1). Label each thermometer with the name of the planet it represents. Let one astronomical unit (AU, the mean distance between the sun and Earth) = 10 cm.

2. Adjust the lamp so that it is on the same plane as the thermometers (Fig. 3.1).

3. Record the starting temperatures for each thermometer in the Data Table.

4. Turn on the lamp. Observe and record the temperature of each thermometer every three minutes for fifteen minutes in the Data Table provided.

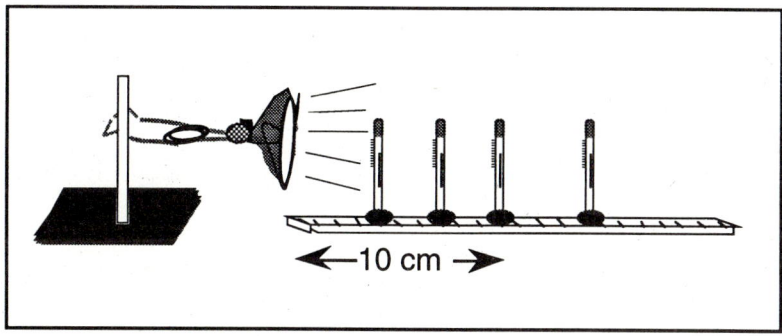

Fig. 3.1

Table 3.1

| PLANET | DISTANCE FROM SUN (AU) |
|--------|------------------------|
| Mercury | 0.38 |
| Venus | 0.72 |
| Earth | 1.00 |
| Mars | 1.52 |

## Data Table

| PLANET | Scale Distance | TEMPERATURE (°C) | | | | | |
|---|---|---|---|---|---|---|---|
| | | Start | 3 min. | 6 min. | 9 min. | 12 min. | 15 min. |
| Mercury | | | | | | | |
| Venus | | | | | | | |
| Earth | | | | | | | |
| Mars | | | | | | | |

QUESTIONS:

1. What happened to the temperatures when the light was turned on? Did the thermometers heat up immediately?

2. Which thermometer showed the greatest rise in temperature? Least rise?

3. Make a graph of the temperature versus distance. Is the relationship linear (a straight line)? Explain the shape of the resulting curve.

4. Why is the hottest time of the day around 3:00 PM even though the sun is at its highest point in the sky at noon?

# *Lab 7: Indirect Measurements*

## INTRODUCTION

In the previous lab, the distances from the sun to various planets in our solar system are given. How do we make measurements of these distances? Unlike objects on Earth, we cannot hold a meter stick up to the planets and measure them. There are, however, indirect methods to make measurements of remote objects. In this lab we use angular diameter to make indirect measurements of objects. This technique can be applied to measuring the diameter of the moon or the height of distant clouds or trees.

Meteorologists apply angular diameter to gauge the height of clouds from radar signals returned as echos by cloud particles. The time it takes for the echo to return provides the distance to the cloud and the clouds angular diameter as seen by the radar can then be used to calculate the height of cloud top. Weather forecasters use this information in determining the hazards represented by storms. The higher the cloud top of a thunderstorm, the more likely it will produce severe weather such as large hail or tornadoes (see section 8).

The easiest way to understand an angular diameter is to look at an example. For instance, it is possible to hold a penny close enough to your eye so that it just blocks out the head of one of your classmates. You have adjusted the position of the penny so that it has the same angular diameter as your classmate's head even though the penny is much smaller. This example shows why angular diameters are useful in making measurements. If the true diameter of the penny is known, the true diameter of the person's head can be mathematically determined without measuring it directly. Many objects in our environment can be measured the same way. The following activity will demonstrate how this principle can be used to determine the height of objects on your campus, the diameter of the moon (The angular diameter of the moon is $0.5°$), and the height of clouds.

## ACTIVITY

OBJECTIVE: The objective of this activity is to learn how angular diameters can be used to make an indirect measurement of the true height of objects.

## MATERIALS:

For each student or group of students:

§ paper plate

§ meter stick

§ construction paper

§ pencil (round ones work best)

§ two push pins

## PROCEDURE:

1. Make a pinhole in the construction paper and attach it to the end of the meter stick with push pins or tacks.

2. Attach the paper plate to the wall in a place so that you can stand exactly 8 meters from it as shown in the following figure. Attach the plate at eye level or a little higher.

3. Point the meter stick at the paper plate and look through the pinhole along the meter stick at the plate.

4. Slide the pencil along the meter stick until the thickness of the pencil just covers the paper plate. The pencil and the plate now have the same angular diameter (Fig. 3.2).

5. Read the distance from the pinhole to the pencil and record it in the Data Table. Add 1 cm to this measurement to account for the fact that the meter stick was not exactly against your eye.

6. Repeat steps 3 through 5 three more times recording each measurement in the Data Table. Then calculate the average of the four measurements and record it in the Data Table as well.

7. Using the meter stick, measure the diameter of the pencil and the distance to the plate and record them in the Data Table. Make sure each measurement is in millimeters.

8. Using the equation below, calculate the diameter of the plate and record it. This is possible because even though the pencil and plate are different sizes, the pencil was moved to the point where it had the same angular diameter as the plate.

$$\text{Diameter of the plate} = \frac{(\text{distance to the plate}) \times (\text{diameter of the pencil})}{(\text{distance to the pencil})}$$

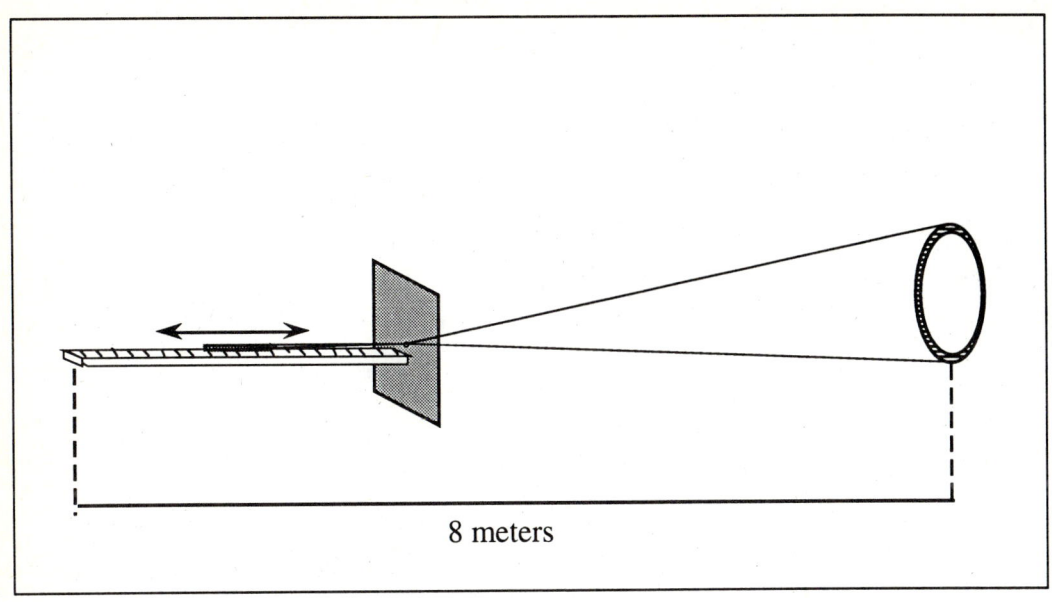

8 meters

Fig. 3.2

9. Measure the true diameter of the plate and record it in the Data Table.
10. Try this activity to determine the height of objects on your school campus. Everything is done the same way except you will measure the height of the object instead of the diameter of a plate. You will need to know the distance to the base of your chosen object; a tree or building, then follow the preceding steps.
11. On a clear night with a full moon this procedure can be used to calculate indirectly the diameter of the moon. To do this you follow the steps above and use the fact that the distance to the moon is ~ 384,401,000,000 mm.

QUESTIONS:
1. Compare the true diameter of the plate with the diameter you calculated. How do the two compare?

2. If the true and calculated diameters are not the same, what reasons could explain the difference?

3. Would this method of determining diameters be helpful in measuring the diameters of the planets? Why?

4. Extra Credit: The full moon appears to be bigger when it is on the horizon than when it is high up in the sky. How could you use the method in this activity to determine whether this appearance is real or an illusion?

**Data Table**

|  | 1 | 2 | 3 | 4 | Average |
|---|---|---|---|---|---|
| Distance from pinhole to pencil (cm) |  |  |  |  |  |
| Diameter of pencil (cm) |  |  |  |  |  |
| Distance to plate (cm) |  |  |  |  |  |
| Diameter of plate (cm) |  |  |  |  |  |
| True diameter of plate (cm) |  |  |  |  |  |

$$\boxed{\phantom{xx}} = \frac{\boxed{\phantom{xx}} \ \text{X} \ \boxed{\phantom{xx}}}{\boxed{\phantom{xx}}}$$

# Lab 8: Surface Heating

## INTRODUCTION

We've all walked across a black top parking lot with our bare feet in the summer. Likewise, it is cooler to wear a white shirt in the hot sun than a black one. But, do you know what causes this and what affect it has on the world around you?

The darker a surface is, the more light it absorbs and the faster it heats up. Conversely, a surface appears light when it has a relatively high albedo (reflectivity) for visible light. The relative ability of different surfaces to absorb light is referred to as absorptivity. The higher a surface's absorptivity the more light it absorbs and is converted to heat. A perfectly black surface absorbs 100 percent of the light that strikes it (albedo = 0). A perfectly white surface reflects 100 percent of the light that strikes it (albedo = 1).

The air above a warm surface is heated as the surface absorbs increasing amounts of light. As the air heats, it becomes less dense and rises. A convection current is generated as cooler air replaces the warmer air and then is subsequently heated as well. This convective motion contributes to the formation of winds.

When adjacent air over the ocean remains much cooler than that over the land, the warm air rises over the land and it is replaced by cooler air from over the ocean. The resulting circulation is referred to as a sea breeze. The temperature contrast from sea to land can be large due the fact that the ocean has a much larger specific heat than the land, and sunlight is absorbed through a greater depth of water. The opposite of a sea-breeze circulation often develops at night when the land cools more quickly than the ocean, and the resultant wind blowing off shore is called a *land breeze*.

## ACTIVITY

OBJECTIVE: The objective of this activity is to investigate the phenomena of differential heating of surfaces.

MATERIALS:
§ reflector lamp with 100-Watt bulb
§ 2 black cups

§ 1 white cup
§ two insulated lids with slits
§ two thermometers
§ ruler or meter stick
§ sand
§ water
§ piece of glass or plexiglass

## PROCEDURE:

1. Slide the thermometers through the slits in the insulated lids so that the bulb of each will be about half way down in the cups.
2. Place the lids on the cups and put the cups side by side, about 10 cm from the lamp as shown in Fig. 3.2.

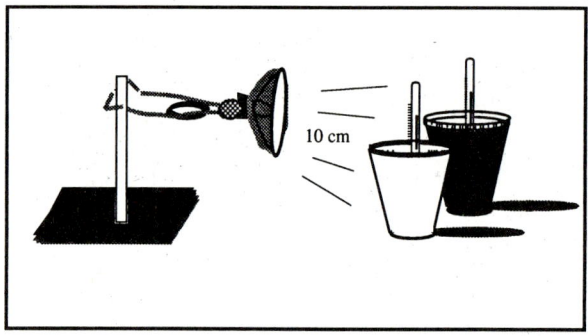

Figure 3.2

3. Record the initial temperature of each cup in the Data Table. In the space below predict what will happen to the temperature in each cup when the light is turned on, and explain your prediction. *Will there be any differences between the cups?*
4. Turn on the light and record the temperatures every two minutes for ten minutes.
5. Record your results in the Data Table below and graph your results on graph paper.
6. Place a piece of glass between the light source and the thermometers and repeat steps 3, 4, and 5. Is there any difference in the rate of temperature change with the glass inserted? Explain briefly below.

7. Fill one black cup with dry sand and the other black cup with wet sand. Repeat steps 1 through 5 above. In the space below predict what will happen to the temperature in each cup when the light is turned on, and explain your prediction. *Will there be any differences between the cups?*

## Data Tables

| | TEMPERATURE (°F) vs TIME (MINUTES) | | | | | |
|---|---|---|---|---|---|---|
| | **0** | **2** | **4** | **6** | **8** | **10** |
| BLACK | | | | | | |
| WHITE | | | | | | |

| | TEMPERATURE (°F) vs TIME (MINUTES) | | | | | |
|---|---|---|---|---|---|---|
| | **0** | **2** | **4** | **6** | **8** | **10** |
| BLACK | | | | | | |
| WHITE | | | | | | |

| | TEMPERATURE (°F) vs TIME (MINUTES) | | | | | |
|---|---|---|---|---|---|---|
| | 0 | 2 | 4 | 6 | 8 | 10 |
| DRY SAND | | | | | | |
| WET SAND | | | | | | |

QUESTIONS:

1. What happened to the temperature measured in the two cups? Was your prediction correct?

2. How, if at all, would the results of this experiment have differed if a silver cup had been used instead of a white one?

3. What would happen if you were to leave the cups under the light for a long period of time, say 24 hours?

4. A glider is an airplane with no engine. In order to stay in the air, glider pilots sometimes look for large paved areas or fields which have been recently plowed. Explain why.

5. What is an advantage of spreading dark sand on snowy roads in locations where it snows a lot in the winter?

# Lab 9:  Solar Radiation

## INTRODUCTION

Among the many consequences of the Earth being round is the fact that its surface is heated differentially by the sun.  The equatorial regions of the Earth receive the maximum amount of solar radiation, whereas the polar regions receive a minimum.

When rays of light strike the surface of an object at a 90° angle, the light is said to be *direct*.  When the angle is anything less than 90°, the light is *slanted*.  Sunlight strikes the Earth along a continuum from most slanted to most direct.  At the equator, the Earth receives nearly direct sunlight and at the poles it receives the most slanted sunlight.  Consequently the equator is always warmer than the poles.

If the Earth were not rotating, this differential heating would set up one global circulation cell with warm air at the equator rising and traveling toward the poles and cold air at the poles sinking and moving toward the equator.  The Earth's rotation makes for a much more complex global circulation pattern that involves several convection cells (this point is returned to in Section 7).

## ACTIVITY

OBJECTIVE: The purpose of this activity is to investigate the heating effect of slanted and direct sunlight.

MATERIALS:
§ 3 identical Celsius thermometers
§ reflector lamp with clamp and 60-watt bulb
§ ring stand with iron ring
§ utility clamp
§ black construction paper
§ stapler
§ books to prop up the thermometer
§ meter stick
§ scissors

## PROCEDURE:

1. Use the black construction paper to cover the bulb of each thermometer as shown. Cut a strip of construction paper 5 cm x 10 cm. Fold the paper in half and staple as shown in Fig. 3.3. Insert the thermometer.

Figure 3.3

2. Prop the thermometers as shown in Fig. 3.4. One thermometer should be vertical, one slanted and one horizontal.

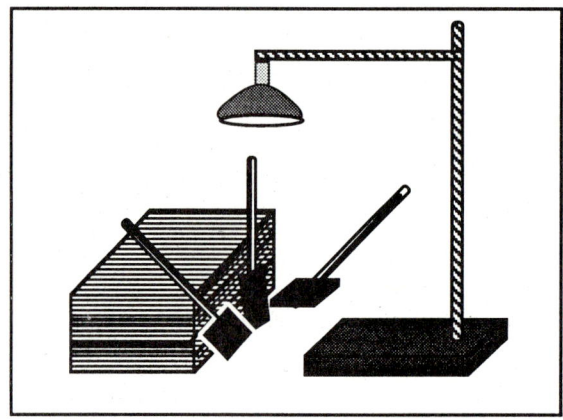

Figure 3.4

3. Adjust the lamp on the ring stand so that the bulb is centered 40 cm above the bulbs of the thermometers.
4. Record the temperature of all three thermometers in the Data Table under the zero column.
5. Turn on the lamp and record temperatures for each thermometer every minute for 15 minutes. Record all temperatures in the Data Table.
6. Using graph paper, make a graph of temperature versus time for each thermometer on the same graph. Using different colored pens or different types of lines (solid vs. dashed) will make the graph for each thermometer easier to distinguish.

## Data Table

| Time(min) | 0 | 1 | 2 | 3 | 4 | 5 | 6 | 7 | 8 | 9 | 10 | 11 | 12 | 13 | 14 | 15 | TotalCh |
|-----------|---|---|---|---|---|---|---|---|---|---|----|----|----|----|----|----|---------|
| Vertical | | | | | | | | | | | | | | | | | |
| Slanted | | | | | | | | | | | | | | | | | |
| Horizontal | | | | | | | | | | | | | | | | | |

QUESTIONS:

1. Which thermometer showed the greatest temperature increase? Why?

2. Why was black construction paper used? Why did we not use a different color construction paper for each thermometer?

3. If you were given a Data Table that listed the average yearly temperatures for cities as you go north from the equator, do you think you would see a trend in the temperatures? If so, what would this trend be and might there be exceptions to the general trend?

# *Lab 10: Local Climate*

## INTRODUCTION

There is a saying, "climate is what you expect and weather is what you get." This saying is accurate in so far as climate is the long term average of weather data. But, climate analysis also deals with extreme weather events. Recent years have brought wild swings in the weather, devastating storms, droughts, and floods. Many people presume these events are the product of a changing climate. But what is climate and what are the factors that control it? The temperature here today depends upon several factors -- seasons, latitude, altitude, proximity to oceans and prevailing weather patterns. In this exercise we will study a year's worth of temperature data and see what we can infer about the factors influencing climate at that location.

Climatological data are conveniently available for many locations in the form of a *Local Climatological Data (LCD) Annual Summary*. *LCDs* are several-page publications available in monthly and annual summary versions for close to 300 National Weather Service offices around the country. The detailed information they report includes temperature averages and extremes, precipitation amounts and extremes, wind values, skycover, snowfall, and heating and cooling degree days. Fig. 3.5 is taken from a *Local Climatological Data (LCD) Annual Summary* for Honolulu, Hawaii.

## ACTIVITY

OBJECTIVES: The purpose of this activity is to investigate one way of comparing the short-term variations of weather and the long-term averages that help define climate. In particular, a year's worth of temperature values for a U.S. location will be compared with the longer-term climatic record at that location in order to

§ contrast differences between weather and climate data
§ use local climate data in planning seasonal activities
§ infer what factors influence climate

## PROCEDURE:

To make this lab as relevant as possible to your location, obtain the *LCD* for a National Weather Service Observing Station near you. One can be obtained from your nearest National Weather Service office or by contacting the National Climatic Data Center, Federal Building, Asheville, NC 28801 (Telephone 704-CLIMATE). Once you have obtained your local LCD answer the questions that follow with respect to this LCD as well as with respect to the LCD provided in Fig. 3.5.

### Annual Summary Of Daily Temperature Data From Honolulu, Hawaii

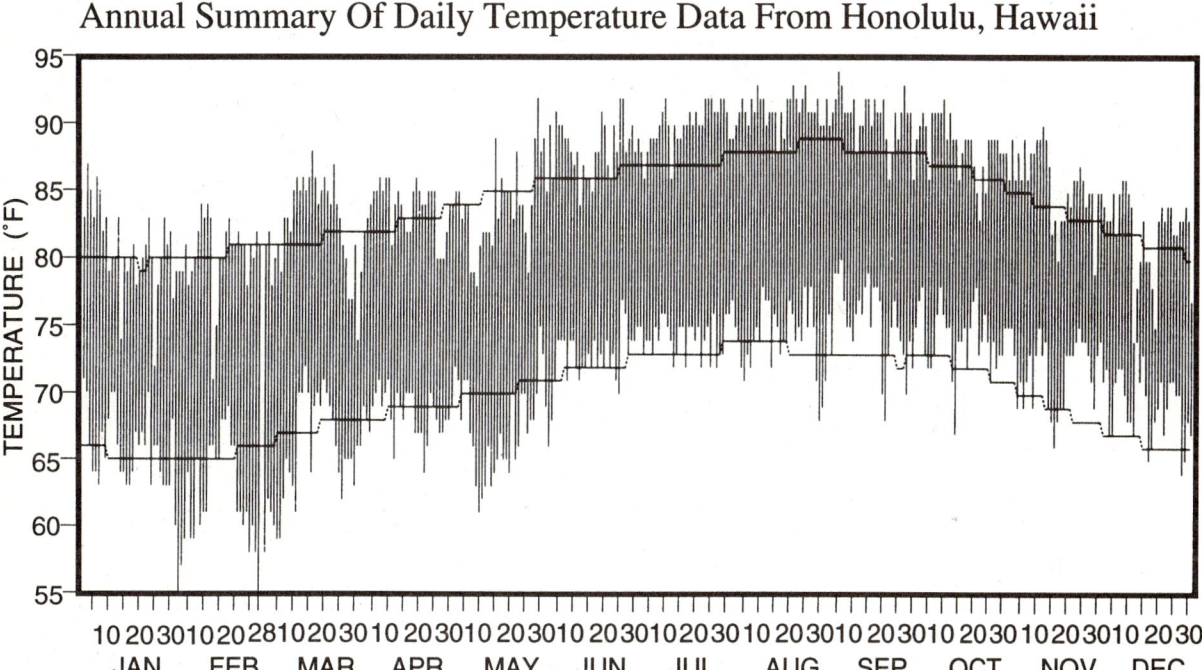

Fig. 3.5

Daily temperatures for 1987 at the Honolulu International Airport National Weather Service observing station are plotted in Fig. 3.5. Each vertical line represents the actual daily temperature high, low, and range for that one day. The two curved lines that stretch across the graph describe normal high and low temperatures throughout the year. Normals are average values typically based on a recent thirty-year period. The upper solid curved line represents daily average maximum temperature, while the lower solid curved line represents minimum average temperature.

49

QUESTIONS:

Refer to the Local Climate Data (LCD) Annual Summary for your location to answer the following questions.

1. In what month or months do the highest daily maximum temperatures *normally* occur at your particular location? When did the highest daily maximum temperature *actually* occur for the year given? What are their values?

2. The *average normal daily temperature* is the average of the *normal* high and low values for the particular day. Draw a curve on the graph to show *normal average daily temperatures* throughout the year.

3. People begin to turn on their air conditioners when the average daily temperature rises above 65 degrees F. For how many months would the air conditioning "season" last if temperatures were completely *normal* throughout the year at your location?

4. How does the daily temperature range of your location compare with that in Fig. 3.5? What factors influence this?

5. How does the seasonal temperature range of your location compare with that in Fig. 3.5? What factors influence this?

6. If temperatures were always *normal* at your location, when would the first fall frost be expected according to the LCD? According to the *actual* data reported in the graph on your LCD, when was the first date it could have occurred?

7. Refering to the Honolulu LCD in Fig. 3.5, would it surprise you to learn that few houses have air conditioning in Honolulu? What factors might reduce the need for airconditioning in Hawaii?

# SECTION 4  THE SPECIAL ROLE OF WATER
## IN THE EARTH'S ATMOSPHERE

Living on a planet whose surface is three quarters ocean, water can easily be taken for granted. However, the contrast between a desert and a jungle illustrates the critical role of water to life on this planet. Water is equally essential to the circulations in the Earth's atmosphere, the formation of storms, and the Earth's climate of itself. It is the peculiar nature of water on the molecular scale that sets water apart from other substances, and clarifies the crucial role of water to the Earth's weather and climate. In this section some of the unusual properties of water will be explored, and ways in which water vapor is measured in the atmosphere will also be discussed.

The water molecule is made up of one oxygen atom and two hydrogen atoms. Each of these atoms are made up of electrons with negative charge surrounding protons with positive charge. The geometrical distribution of these charges in the water molecule is such that on the side of the water molecule where the hydrogen atoms are attached there is slightly more positive charge than negative charge; conversely on the oxygen atom side of the water molecule there is slightly more negative charge (See Fig. 4.1). Because of

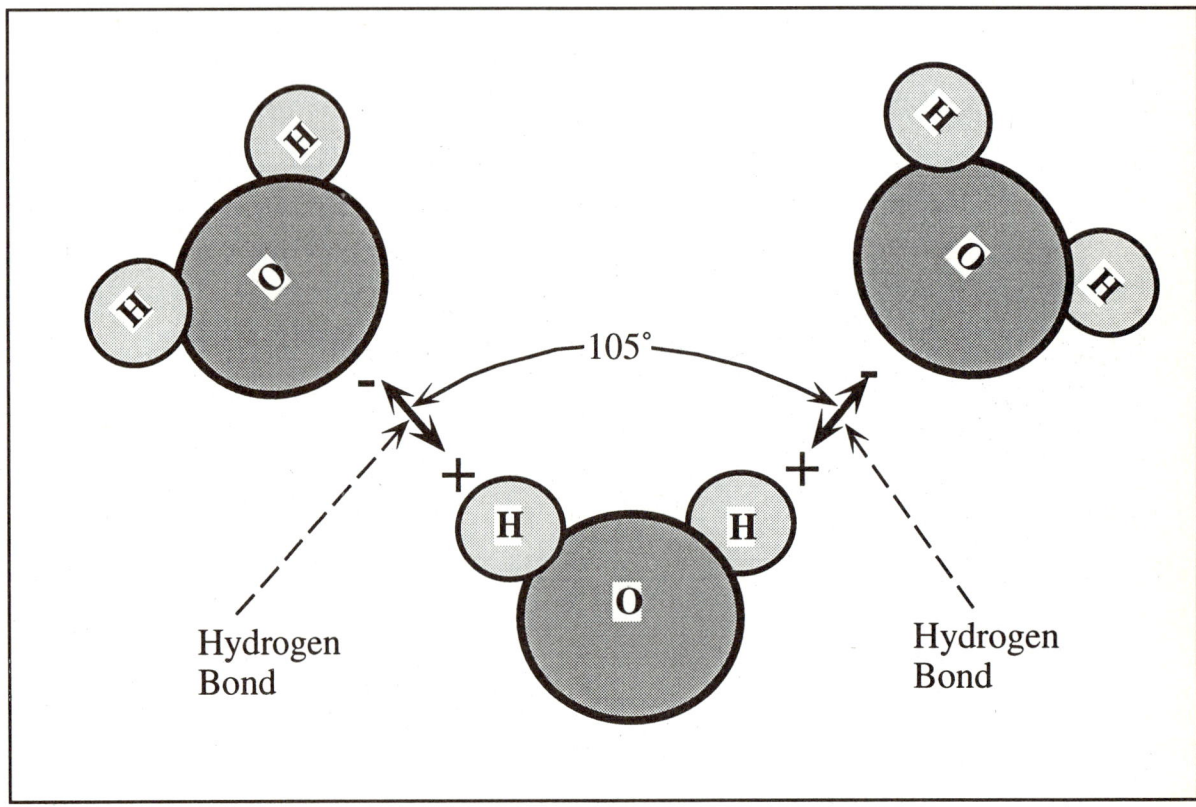

Figure 4.1 Bonding of water molecules

this distribution, water is called a *polar* molecule. Therefore, when two water molecules approach, the positive side of one will be attracted to the negative side of the other. The resultant bond that forms between the two water molecules is called a *hydrogen bond*. When a hydrogen bond forms heat is released into the air, and conversely when a hydrogen bond breaks heat is consumed. Hydrogen bonds are relatively strong bonds, consequently the heat is associated with the bonds is large.

In liquid water there are millions of hydrogen bonds that rapidly break and reform, giving liquid water its fluid character. Consequently, liquid water is sometimes referred to as a *pseudo-crystalline* substance. It is the energy used in breaking hydrogen bonds that accounts for the large amount of heat required to heat a pan of water. Thus water is said to have a high *heat capacity.* Heat capacity is the ratio of heat absorbed by a substance to its corresponding rise in temperature. Specific heat is a measure of the heat capacity of a substance. (Recall that it requires one calorie to raise the temperature of water by one degree Celcius.) The oceans are, therefore, great moderators of the climate. In the solid or ice phase, water molecules are tightly bonded to each other in a regular crystal lattice. The geometry of the lattice gives rise to the beautiful hexagonal and dendritic snowflakes. In the vapor phase all of the hydrogen bonds are broken and the water molecules reside in a solitary gaseous phase.

There are several ways in which water molecules can change phase. In each case hydrogen bonds are either formed or broken, and heat is released or consumed. The heat associated with such changes of phase is referred to as *latent heat*, since it remains "hidden" until the phase change occurs. The magnitudes of latent heats vary with temperature, but at zero degrees Celcius, the latent heat of melting is 80 calories per gram, and for evaporation it is 580 calories per gram. These relationships are summarized below and in Fig. 4.2:

*(evaporation)* liquid to vapor - hydrogen bonds broken, 580 cal/g latent heat consumed
*(condensation)* vapor to liquid - hydrogen bonds formed, 580 cal/g latent heat released
*(melting)* ice to liquid - hydrogen bonds broken, 80 cal/g latent heat consumed
*(freezing)* liquid to ice - hydrogen bonds formed, 80 cal/g latent heat released
*(sublimation)* ice to vapor - hydrogen bonds broken, 677 cal/g latent heat consumed
*(deposition)* vapor to ice - hydrogen bonds formed, 677 cal/g latent heat released

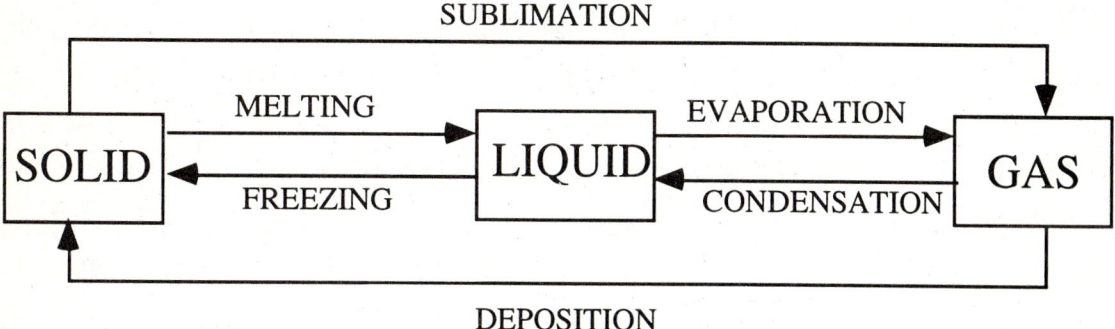

Figure 4.2 Changes of phase in water

Here are several common examples from daily life. When stepping from a shower into a dry room your wet skin feels cold. Heat is required to break all of the hydrogen bonds when water evaporates, and your skin provides much of that heat. Thus, until all the liquid water evaporates (or is removed by towel), your skin will continue to feel the cooling effect of evaporation. Conversely, when water droplets form on the outside of a glass of cold beer, condensation occurs and heat is released as the hydrogen bonds form, thus warming the beer. When clouds form, water molecules condense onto liquid cloud droplets, and heat is released into the cloudy air as the hydrogen bonds reform. This heat adds to the buoyancy of the rising cloud. It is the energy invested in the hydrogen bond that helps fuel thunderstorms, hurricanes and acts to moderate the climate of the Earth. More will be said about these topics in Sections 6 and 8.

### *Measuring water vapor in the atmosphere*

Having established the importance of water in transferring heat, how does one measure water vapor in the atmosphere? When air contains no water molecules, the vapor pressure is equal to zero; *vapor pressure* is the contribution to the total pressure by water vapor molecules. Given sufficient time, and a fixed temperature, the vapor pressure over a flat surface of water will reach equilibrium. In such a state, the rate at which water molecules leave the water surface equals the rate at which they enter the surface. At this point the air is said to be *saturated* and the *relative humidity* equals 100 percent. The relative humidity is the ratio of the actual vapor pressure to the vapor pressure at saturation expressed in percent. In some circumstances, the quantity of water vapor in the air exceeds the saturation value, the air is *supersaturated*, and the relative humidity is greater than 100 percent.

The saturation vapor pressure depends only on the temperature and increases rapidly as the temperature increases. An illustration of the rapid increase in saturation vapor pressure with temperature is the fact that air at 40°C with relative humidity of 10% contains more water vapor than air at -10°C with a relative humidity of 100%.

Air over a surface of ice may also reach water-vapor equilibrium and become saturated with respect to ice. The saturation vapor pressure with respect to ice is less than the saturation pressure with respect to water because water molecules are more completely bonded to ice. Therefore, in a

54

cloud composed of both ice crystals and liquid droplets, the ice will grow rapidly at the expense of the droplets. This point is important for the formation of precipitation and will be returned to in Section 5.

Early scientists noticed that certain natural fibers are sensitive to the amount of water vapor in the air. Instruments called *hygrometers* were constructed using animal hairs to measure the humidity of the air. Another way to measure water vapor is the *psychrometer*. A psychrometer employs two thermometers, one measuring the air temperature and the second measuring the wet-bulb temperature. The latter is obtained by covering a thermometer bulb with a wet muslin wick and ventilating it - as water evaporates, the temperature decreases. The lowest temperature attained is the *wet-bulb temperature*. Knowing the air temperature and the wet-bulb temperature, it is possible to determine from psychrometric tables the relative humidity or other measures of atmospheric humidity, such as the *dew-point temperature*. The dew point temperature is the temperature at which water droplets (or dew) form on a cooled, clean plate. This temperature is sometimes confused with the wet-bulb temperature. The wet-bulb temperature is actually a bit higher than the dew-point temperature because some of the water vapor from the wick moistens the air as the reading is taken.

Although many of the unique properties of water are discussed in this section several additional peculiarities of water have been deferred the section on clouds and precipitation (section 6). Implications for the large latent heats associated with water's phase changes are discussed further in section 6 and in section 8 on thunderstorms and hurricanes.

# LAB 13: EVAPORATION AND LATENT HEAT

Of all the basic molecules on Earth, one of the most important and most common is water. Water is the compound that has made the Earth the only life-bearing planet in the solar system. Much of the Earth's uniqueness may be attributed to the presence of water and its unusual properties.

There is ample evidence in the open environment that water evaporates and enters the atmosphere. Surfaces dry after a rain. Water puddles gradually disappear. Clothes dry in the open air. But is it only water that is involved in the process?

## ACTIVITY

OBJECTIVES: The objective of this activity is to observe the impact of the energy (latent heat) required to break hydrogen bonds when water molecules evaporate into the atmosphere and to probe ways in which latent heat makes life more comfortable for humans.

MATERIALS:
§ two thermometers
§ one plastic (or plastic coated paper) plate
§ one sheet of clear plastic food wrap large enough to cover the plate
§ about 0.5 liter of water

PROCEDURE:
1. Bring the water to room temperature by either leaving a closed jar of water in the room for several hours or by mixing warm and cold water while checking its temperature with a thermometer.
2. Place the paper plate on the table and put one of the thermometers facing up in the middle of the plate. Place the other thermometer near the plate on the table in a position so that both thermometers can be easily read (Fig. 4.3). Start the activity with the water at room temperature. Pour a shallow layer of the water into the plate.

3. Read the thermometers every two minutes and record your observations. Continue until both air and water temperatures attain steady values. Record your observations in the Data Table and make a graph of temperature versus time for each thermometer on the same graph.

Fig. 4.3

**Data Table**

|  | 2 MIN. | 4 MIN. | 6 MIN. | 8 MIN. | 10 MIN | 12 MIN | 14 MIN |
|---|---|---|---|---|---|---|---|
| AIR | ° | ° | ° | ° | ° | ° | ° |
| WATER | ° | ° | ° | ° | ° | ° | ° |

QUESTIONS:

1. What happens to the air and water temperatures? Explain what you think is happening to produce what you have observed?

2. What would eventually happen to the water in the open plate if left long enough?

Cover the plate with a clear plastic sheet or food wrap stretched so it does not touch the water and continue to observe temperatures.

3. What happens to the water temperature now? How does it compare with the room temperature? Graph your observations.

4. What will happen to the amount of water in the covered plate, even if left overnight or longer? What are the implications for a greenhouse?

5. People living in hot and dry climates have long used porous unglazed pottery to store drinking water. Why?

6.  An economical air conditioner in dry climates is what is sometimes called a "swamp cooler". These devices are designed to cool air by evaporating water. Based on what you have observed in this activity, describe how they are designed.

7. On a global average, a layer of water about 100 centimeters in depth is evaporated to the air each year. About 600 calories of energy are required to evaporate one cubic center of water at Earth-surface temperatures. Approximately how many calories of energy are transported to the atmosphere above each square centimeter of the Earth's surface each year?

8. What happens to this energy when atmospheric water vapor condenses to form clouds?

9. Look at the satellite view of the Earth on the cover of this book. Does the distribution of clouds in the image suggest anything about how clouds act to redistribute heat over the globe?

# Lab 14: Measuring Moisture in the Air

## INTRODUCTION

We are all familiar with the hot, humid summer afternoons that plague the East Coast. On these days, there is an abundance of water vapor in the air. The amount of water vapor the air can hold depends upon the temperature of the air. Warm air can hold much more water vapor than cold air. When warm humid air rises in thermals it expands, cools, and quickly becomes saturated. When this happens, water condenses out in the form of cloud droplets and the result is cumulus clouds so prevalent on warm summer afternoons.

### Dew Point

One way to measure the water vapor content of the atmosphere is by the dew point temperature. The dew point temperature is the temperature at which water droplets first form on a cooled, clean surface. If water vapor does condense on a surface, this indicates that the temperature of the surface is below the dew point temperature. Higher dew point temperatures indicate a higher moisture content; lower dew point temperatures indicate a lower moisture content. When the dew point temperature and the air temperature are the same, the relative humidity is 100% and the air is said to be saturated.

## ACTIVITY

OBJECTIVE: The purpose of this activity is to measure the amount of moisture in the atmosphere by observing the formation of dew.

MATERIALS:
§ water and ice cubes
§ thermometer
§ tin can with a shiny, clean surface

PROCEDURE:
1. Measure and record the temperature of the air on the data sheet.
2. Fill the can 3/4 full of water. Measure the temperature of the water as shown in Fig. 4.4. Record on the data sheet.

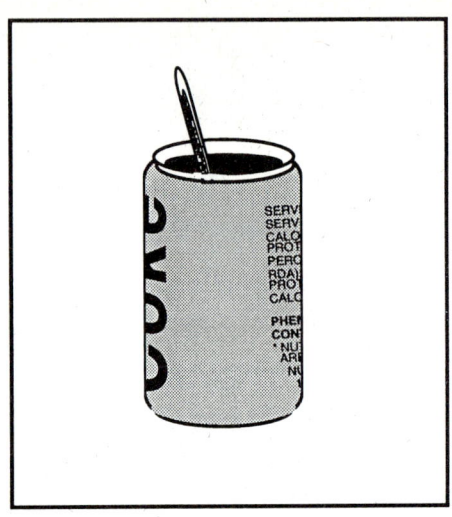

Figure 4.4

3. Slowly add small pieces of ice to the can while carefully stirring constantly with the thermometer. Watch the outside of the can closely for the first sign of condensation.
4. When the first condensation forms, immediately record the temperature of the water in the can under the column "Dew Point" in the Data Table.
5. Use a conversion table to convert the dew point temperature and temperature that you measured to relative humidity.
6. Repeat steps 1-5 outside.

* If condensation does not form before the ice cubes melt, add more ice cubes and continue stirring until it does form.

QUESTIONS:
1. When air rises, it expands and cools ~1 degree Celsius for every 100 m of altitude. If the air in the room were to rise, approximately at what height would the moisture begin to condense and form clouds? Show your work.

2. Repeat #2 for the air outside. Are there any clouds in the sky near this height?

## Data Tables

### *Inside*

| Trial | Air Temp. (°C) | Dew Point Temp. (°C) | Average Dew Point Temp. |
|-------|----------------|----------------------|-------------------------|
| 1 | | | |
| 2 | | | |
| 3 | | | |
| | | | |

### *Outside*

| Trial | Air Temp. (°C) | Dew Point Temp. (°C) | Average Dew Point Temp. |
|-------|----------------|----------------------|-------------------------|
| 1 | | | |
| 2 | | | |
| 3 | | | |
| | | | |

3. What atmospheric conditions would allow water vapor from the air to condense on your skin?

4. How does the water vapor content of the air differ from inside to outside? What accounts for this difference?

# *Lab 15: Wet-Bulb Temperature*

## INTRODUCTION

We experience the cooling associated with the latent heat of evaporation in our daily lives. Each time we step out of the shower, evaporation provides the largest part of the chill we feel. More chill is experienced when the air in the room is dry than if the air is humid. Another common example is the cool feeling of grass on bare feet, also due to the evaporation of water from the grass blades.

We can employ the cooling associated with evaporation to measure humidity in the atmosphere. This experiment will employ a sling psychrometer to measure the amount of moisture in the atmosphere by observing how far the latent heat of evaporation can depress the temperature of a "wet" thermometer.

## ACTIVITY

OBJECTIVE: The purpose of this activity is to measure the moisture in the atmosphere through measuring the affect of the latent heat of evaporation on a wet thermometer.

MATERIALS:
§ water
§ sling psychrometer

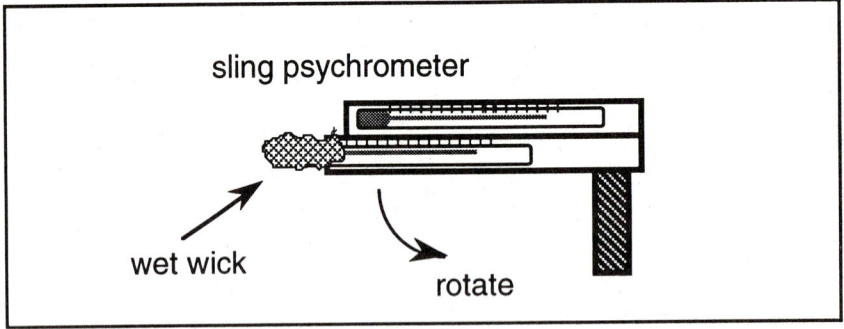

Fig. 4.5

## PROCEDURE:

1. Using the sling psychrometer, measure and record the indoor air temperature (Fig. 4.5).
2. Wet the muslin of the psychrometer with water. Sling the psychrometer over your head for one minute. Make sure you have checked to make sure that nothing is in the way!
3. After a one-minute period, read the wet bulb temperature. Spin the psychrometer around for another thirty seconds. If the wet bulb temperature has not changed, record the temperature in the Data Table. If it has changed, sling the psychrometers for thirty-second periods until you get two readings that are the same. Record this temperature in the Data Table.
4. Repeat the above procedure outside.
5. Determine the relative humidity inside and out by using the attached table.

### Data Table

| Location | Time | Wet Bulb Temp. (°C) | Dry Bulb Temp. (°C) | Difference (°C) | Relative Humidity |
|---|---|---|---|---|---|
| Inside | | | | | |
| Outside | | | | | |

## QUESTIONS:

1. Compare the relative humidities inside and out. What accounts for the difference?

2. Why do you sling the psychrometer until you have two alike readings?

3. If the temperature were to rise and the amount of water vapor were to remain constant, what would happen to the wet-bulb temperature and the relative humidity?

63

## Table to Determine Relative Humidity

Dry Bulb Temperature (°C)

Difference Between Dry Bulb and Wet Bulb Temperature (°C)

| Diff | 5 | 6 | 7 | 8 | 9 | 10 | 11 | 12 | 13 | 14 | 15 | 16 | 17 | 18 | 19 | 20 | 21 | 22 | 23 | 24 | 25 | 26 | 27 | 28 | 29 | 30 | 31 | 32 | 33 | 34 | 35 |
|---|---|---|---|---|---|---|---|---|---|---|---|---|---|---|---|---|---|---|---|---|---|---|---|---|---|---|---|---|---|---|---|
| 1 | 86 | 86 | 87 | 87 | 88 | 88 | 89 | 89 | 90 | 90 | 90 | 90 | 90 | 91 | 91 | 91 | 92 | 92 | 92 | 92 | 92 | 92 | 93 | 93 | 93 | 93 | 93 | 93 | 93 | 93 | 94 |
| 2 | 72 | 73 | 74 | 75 | 76 | 77 | 78 | 78 | 79 | 79 | 80 | 81 | 81 | 82 | 82 | 83 | 83 | 83 | 84 | 84 | 84 | 85 | 85 | 85 | 86 | 86 | 86 | 86 | 87 | 87 | 87 |
| 3 | 58 | 60 | 62 | 63 | 64 | 66 | 67 | 68 | 69 | 70 | 71 | 71 | 72 | 73 | 74 | 74 | 75 | 76 | 76 | 77 | 77 | 78 | 78 | 78 | 79 | 79 | 80 | 80 | 80 | 81 | 81 |
| 4 | 45 | 48 | 50 | 51 | 53 | 55 | 56 | 58 | 59 | 60 | 61 | 63 | 64 | 65 | 65 | 66 | 67 | 68 | 69 | 69 | 70 | 71 | 71 | 72 | 72 | 73 | 73 | 74 | 74 | 75 | 75 |
| 5 | 33 | 35 | 38 | 40 | 42 | 44 | 46 | 48 | 50 | 51 | 53 | 54 | 55 | 57 | 58 | 59 | 60 | 61 | 62 | 62 | 63 | 64 | 65 | 65 | 66 | 67 | 67 | 68 | 68 | 69 | 69 |
| 6 | 20 | 24 | 26 | 29 | 32 | 34 | 36 | 39 | 41 | 42 | 44 | 46 | 47 | 49 | 50 | 51 | 53 | 54 | 55 | 56 | 57 | 58 | 58 | 59 | 60 | 61 | 61 | 62 | 63 | 63 | 64 |
| 7 | 7 | 11 | 15 | 19 | 22 | 24 | 27 | 29 | 32 | 34 | 36 | 38 | 40 | 41 | 43 | 44 | 46 | 47 | 48 | 49 | 50 | 51 | 52 | 53 | 54 | 55 | 56 | 57 | 57 | 58 | 59 |
| 8 |  |  |  | 8 | 12 | 15 | 18 | 21 | 23 | 26 | 27 | 30 | 32 | 34 | 36 | 37 | 39 | 40 | 42 | 43 | 44 | 46 | 47 | 48 | 49 | 50 | 51 | 51 | 52 | 53 | 54 |
| 9 |  |  |  |  |  | 6 | 9 | 12 | 15 | 18 | 20 | 23 | 25 | 27 | 29 | 31 | 32 | 34 | 36 | 37 | 39 | 40 | 41 | 42 | 43 | 44 | 45 | 46 | 47 | 48 | 49 |
| 10 |  |  |  |  |  |  |  |  | 7 | 10 | 13 | 15 | 18 | 20 | 22 | 24 | 26 | 28 | 30 | 31 | 33 | 34 | 36 | 37 | 38 | 39 | 40 | 41 | 42 | 43 | 44 |
| 11 |  |  |  |  |  |  |  |  |  |  | 6 | 8 | 11 | 14 | 16 | 18 | 20 | 22 | 24 | 26 | 28 | 29 | 31 | 32 | 33 | 35 | 36 | 37 | 38 | 39 | 40 |
| 12 |  |  |  |  |  |  |  |  |  |  |  |  |  | 7 | 10 | 12 | 14 | 17 | 19 | 20 | 22 | 24 | 26 | 27 | 28 | 30 | 31 | 32 | 33 | 35 | 36 |

Rainbows, halos, glories, and similar optical phenomena observed in the atmosphere provide for some of the most breathtaking spectacles in nature. It is not surprising that religious writings have interpreted the sighting of such phenomena as omens variously portending prosperity, death, or war. In no way diminishing their splendor, atmospheric optics are generally the result of a combination of one or more of the following four physical mechanisms: scattering, refraction, reflection, and diffraction.

_Scattering_ is the process by which small particles diffuse a portion of the incident radiation in all directions. The amount of light that is scattered depends upon the size of the particle. Scattering of sunlight by air molecules is very sensitive to the wavelength of the light with the shorter blue wavelengths scattered almost four times more efficiently than the longer red wavelengths. This accounts for the blue sky we see. Scattering by air molecules is referred to as _Rayleigh scattering_, after Lord Rayleigh who first described it. In the absence of Rayleigh scattering the sky would appear black, as it does on the moon. As the sun approaches the horizon, the path length of the sunlight through the air increases and more and more of the blue light has been scattered away by the intervening air, thus the sun appears increasingly red.

For larger particles such as cloud droplets, haze, and smoke, the scattering falls into what is called the *Mie scattering* regime (named after Gustav Mie who developed the more general theory of light scattering) in which there is less dependence on wavelength, rendering the scattered light neutral or whitish in color. This regime accounts for the white appearance of clouds when the sun is high.

Another physical mechanism that results in atmospheric optical phenomena is *refraction.* Refraction is the bending of light as it passes from one medium to another (Figure 5.1). Refraction can also occur within a given medium when density differences cause the light to bend gradually. Refraction results from the fact that the speed of light varies according to the property of the medium through which it passes.

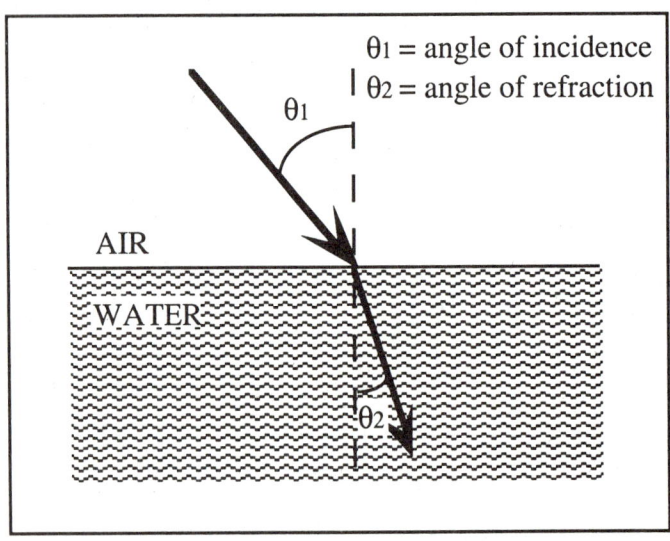

Figure 5.1 Refraction of sunlight by water

The amount of refraction depends upon the density of the material, the angle at which the light enters the material, and the wavelength of the light. As a result of the wavelength dependence, white light is separated into a spectrum of colors after it passes through a glass prism. Red light is refracted least due to its large wavelength, and blue light is refracted most

66

due to its short wavelength.

A common physical mechanism that produces interesting optical effects is *reflection*. When reflection occurs an object or ray of light bounces off a surface at the same angle at which it strikes the surface. This is best stated in the *Law of Reflection*: the angle of incidence is equal to the angle of reflection (Figure 5.2). For instance, when light strikes a mirror, it is reflected in the same way a ball is reflected after striking a wall. Surfaces of discontinuity, such as air- water interfaces, result in a partial reflection of the incoming radiation.

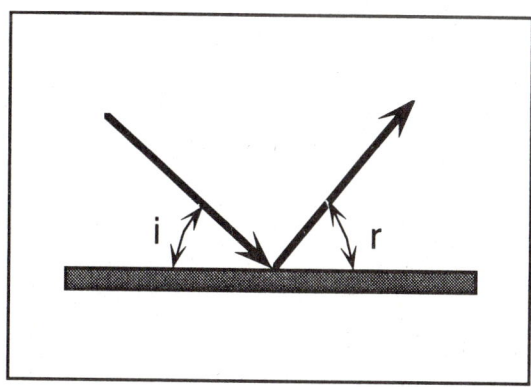

Figure 5.2  Law of reflection (i = r)

The fourth physical mechanism mentioned is *diffraction*. Diffraction occurs when light bends along the boundary of an object. Diffracted light from many uniformly sized cloud droplets can interfere in such a way as to produce a pattern of colored light. A region of bright color is then the place where light waves of a certain wavelength have crests that arrive simultaneously. The rainbow colored pattern seen in modern compact disks is an example of diffraction.

### Rainbows

*Rainbows* are optical phenomena created as a result of both refraction and reflection of sunlight by raindrops. As sunlight enters a raindrop, it is refracted once then reflected off the backside of the droplet and refracted again before the light leaves the drop (Figure 5.3) The result is known as a primary rainbow because light is reflected only once. Rainbows can be seen when an observer has his back to the sun and is facing illuminated rain.

As explained earlier, refraction separates light according to the wavelength of the light. When a rainbow is formed, the light with a longer wavelength (red) appears along the outer edge of the rainbow and the light with a shorter wavelength (blue) is along the inside. An acronym that is commonly used to remember the color order of a rainbow is ROY G. BIV (red, orange, yellow, green, blue, indigo and violet).

Often times a second, lighter rainbow is observed next to the primary rainbow. This is known as a secondary rainbow because the light in the

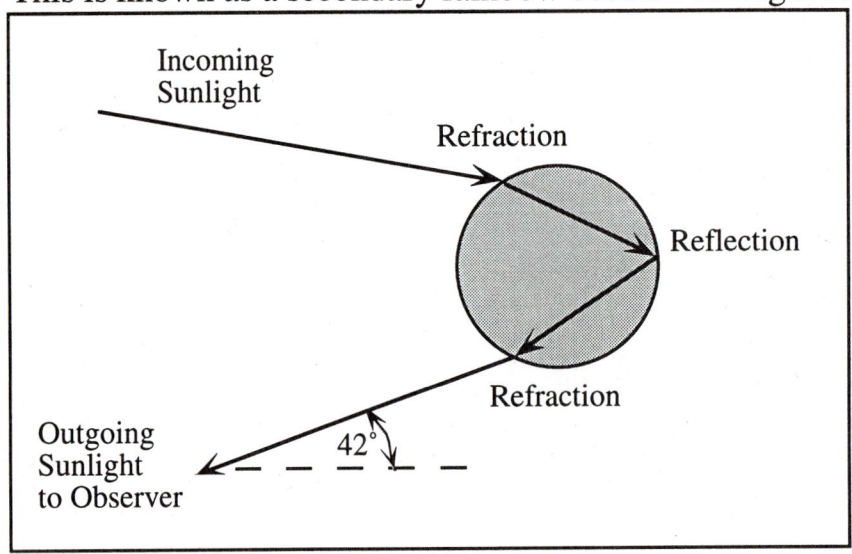

Figure 5.3 Primary rainbow

raindrops is reflected two times, instead of only once (Figure 5.4). The order of the colors of the secondary rainbow are reversed, with the blue on the outside and the red on the inside.

Another optical phenomena frequently observed is the *halo*. The word halo is a general term referring to any ring or arc that is created as a result of reflection and/or refraction of light by ice crystals. Most commonly, a halo forms around the sun or moon when light rays are refracted by ice crystals in cirrostratus clouds.

Halos are often observed as whitish rings around the sun or moon. A distinct rainbow spectrum is sometimes be observed around the sun. When color is present in a halo, red is closest to the sun, followed by yellow, green and blue on the outside. The colors of a halo are not usually as vivid as they are in a rainbow, and thus are often overlooked. The most common halo occurs at a radial distance of 22.5° from the sun, and is produced as a result

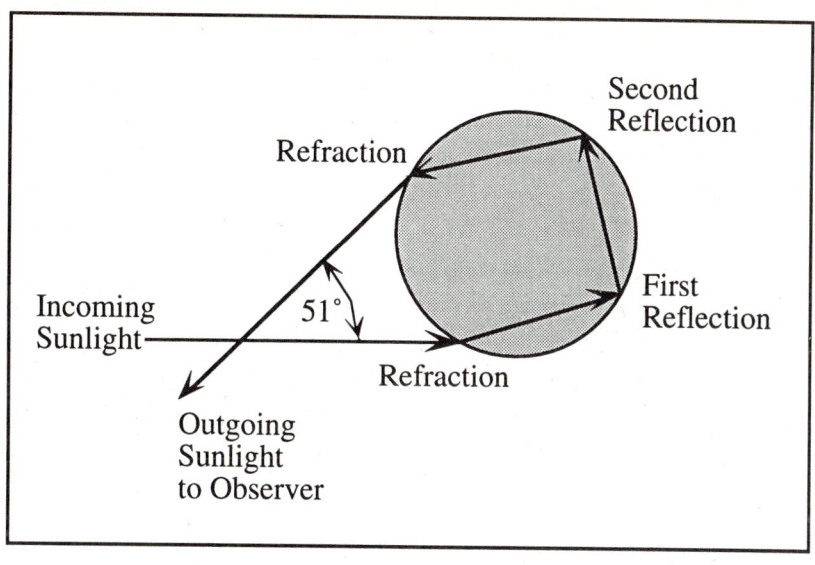

Figure 5.4 Secondary rainbow

of refraction of sunlight through the sides of hexagonal, pencil-shaped ice crystals that are oriented randomly.

The presence of altocumulus and cirrocumulus clouds composed of uniform sized cloud droplets allow light waves from the sun or moon to bend around them and thus create an interference pattern we call a *corona*. Corona are very common around the moon and are often confused with halos, but are actually the result of a very different physical mechanism, diffraction. The radius of a corona, typically only a few degrees, is much smaller than common halos. Colorful coronas around the sun are sometimes called Bishop's rings and differ from halos in their color order. Red is present on the outer edge, then fades to a blue or almost white along the inside. In good examples of Bishop's rings several concentric rainbow-colored rings can appear. The presence of color in a corona is the result of interference of light waves.

*Glories* are another type of atmospheric optical phenomenon that many people overlook. In order to view a glory, the observer must be in bright sunshine and above clouds or fog. Therefore, a common place for viewing a glory is from an airplane or mountain ridge. A glory appears as a ring of color located around the shadow of the observer and has the reverse color order of a corona. The glory forms through a complex combination of diffraction and reflection of light by the cloud by droplet. Cloud droplets that produce glories are of uniform size, as is the case for coronas.

# Lab 16: Bending Light

## INTRODUCTION

In Fig. 5.1, a beam of light strikes a water surface at an angle. At the interface between the water and the air, part of the beam is reflected and part enters the glass. Notice that the light entering the glass bends at an angle. This bending is called refraction, and it occurs whenever light crosses, at an angle, the boundary between two media of different densities.

In Fig. 5.1, light is traveling from a medium of low density to one of high density (air to glass). As the light enters the glass, it slows down and refracts toward normal. As $n_2$ (index of refraction) in the second medium increases, the angle $\theta_2$ (as measured from the normal) decreases.

Willebrord Snell (1591-1626) is credited with the discovery of the relationships involved in refraction. He found that the change in velocity is analytically related to the angles formed at the boundary and the properties of the media through which the light is traveling. That is,

$$V_2/V_1 = \sin\theta_2/\sin\theta_1 = \text{constant}$$

$V_1$ = velocity of light in first medium
$V_2$ = velocity of light in second medium

The constant is a dimensionless number called the index of refraction and is represented by the letter "n." From the above equation, we get the widely used form of Snell's Law:

$$n_1\sin\theta_1 = n_2\sin\theta_2 .$$

$n_1$ = index of refraction in first medium
$n_2$ = index of refraction in second medium
$\theta_1$ = angle of incidence
$\theta_2$ = angle of refraction

This is the equation you will use to solve the problems in the following activity.

70

# ACTIVITY

OBJECTIVE: The purpose of this activity is to draw the path a light ray follows as it travels through air, fresh water and salt water. You will then measure the angles formed and determine the index of refraction for fresh water and salt water.

MATERIALS:
§ glass or plexiglass rectangular box
§ straight pins
§ cardboard
§ white paper
§ protractor with straight edge
§ salt water (50 grams of table salt to 1 liter of water)

PROCEDURE:
1. Secure a piece of paper to the cardboard and set it down on a table. Place the glass box on the center of the paper. Trace the outline of the box. Remove the box from the paper (see Fig. 5.5).
2. Find the midpoint on the back of the outline and mark its position (next to the line, not on it). Label this point B.
3. Draw a dotted line through Point B that is perpendicular to the outline of the box. This represents the normal to the back face of the box.
4. Draw another line, two centimeters in length, that extends out from Point B at a 45-degree angle. Label the point at the end of this line "A" and the angle "$\theta_1$".
5. Stick a straight pin into the paper at Point A and another one at Point B. Fill the glass box with fresh water and place it back onto the paper inside the outline.
6. From the front of the glass box, position yourself so that your line of sight is just above the table top. You should have a clear view of the pins on the other side. Move left or right until the pins at Points A and B appear evenly aligned. (You may close one eye, but do not tilt your head.)
7. Once aligned, mark a third point, "C," on the front side as close to the glass box as possible. Stick a straight pin into the paper at Point C. All three pins should appear to be aligned.

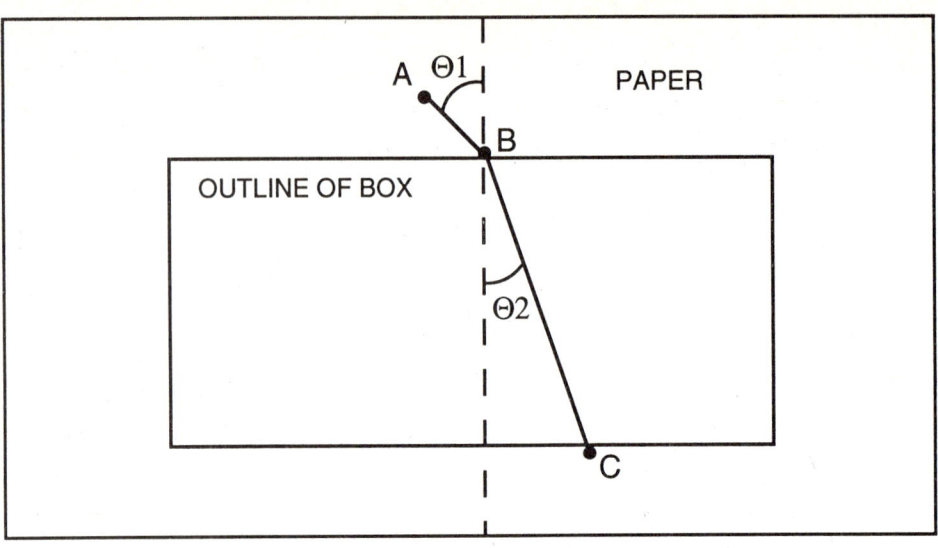

Fig. 5.5

8. Carefully remove the glass box from the paper. Draw a line from Point B to Point C. Label the angle this line makes with the normal, "$\theta_2$".

9. Measure the angle with a protractor.

10. Using Snell's Law, calculate the index of refraction for fresh water.

11. Fill the glass box with salt water and place it back over the outline. Repeat the steps above. Label your new point of alignment, "D," and the new angle, "$\theta_2^2$."

12. Calculate the index of refraction for salt water.

QUESTIONS:

1. The accepted value for the index of refraction in fresh water is 1.33. How does your measured value compare with the accepted value? Express your answer in terms of percent error.

$$\% \text{ Error} = \frac{\text{Difference between measured \& accepted value}}{\text{accepted value}} = 100\%$$

% Error = _____ %

72

2. Why is there no single accepted value for the index of refraction of seawater?

3. In the experiment, we did not take into account what happens to the light as it passes through the glass. In terms of speed and direction, how does the light behave as it travels from the air to the glass? From the glass to the water? From the water to the glass? And, finally, from the glass to the air again? Fill in Table 5.1 below.

Table 5.1

|  | SPEED | DIRECTION |
|---|---|---|
| AIR TO GLASS: |  |  |
| GLASS TO WATER: |  |  |
| WATER TO GLASS: |  |  |
| GLASS TO AIR: |  |  |

# Lab 17: Creating Rainbows

## INTRODUCTION

It's always inspiring to see a rainbow after a rain shower. By now, most people realize that finding a pot of gold at the end of a rainbow is nothing more than a myth. Rainbows are the result of sunlight being refracted and reflected through water droplets. This activity will let you explore the formation of a rainbow and what helps or hinders the visibility of rainbows.

## ACTIVITY

OBJECTIVE: The objective of this activity is to create and investigate a rainbow.

MATERIALS:
§ large glass beaker
§ water
§ cardboard
§ scissors
§ flashlight
§ milk
§ stirring rod

PROCEDURE:
1. Fill the glass beaker with water.
2. Cut a circle in the cardboard that is approximately the size of the beaker.
3. Take the beaker and cardboard into a dark room. Shine the flashlight through the hole in the cardboard into the beaker (See Fig. 5.6). **Record your observations.**
4. Vary the distance between the cardboard and the beaker. **Record your observations.**
5. Add a few drops of milk (can use non-fat dry milk or coffee creamer for this purpose) to the beaker and stir well. Repeat the above procedure. **Record any changes in your observations.**
6. With the milk added to the water, view the light through the beaker by looking opposite the light source. **Record your observations.**

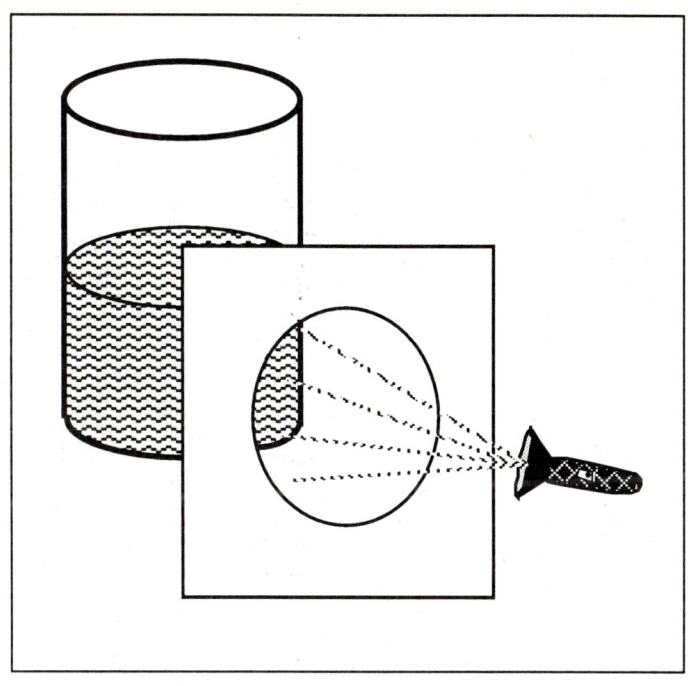

Figure 5.6

QUESTIONS:

1. Why did a rainbow appear on the cardboard?

2. How did the distance between the beaker and the cardboard effect the image? Why is this so?

3. After the milk was added to the water how does the color of the light of the flashlight seen through the beaker compare with the color of the light scattered by the milk in the beaker? What atmospheric optical phenomena do these observations relate to?

4. How, if at all, is this image different from the one you see in the sky?

5. Can a rainbow be produced from moonlight? Why or why not?

6. Why don't you see a rainbow near noon in the summer?

7. Explain why refraction adds to the length of the day.

8. If you were looking for a rainbow in the morning, which direction would you look? Why that direction?

9. Pick a sunny day and set out a sprinkler to make your own rainbow. When you see the bow where is the sun relative to the direction in which you are looking? How does the droplet size in the sprinkler spray affect the appearence of the rainbow?

# LAB 18: INVESTIGATING REFRACTION

## ACTIVITY

OBJECTIVE:  The purpose of this activity is to observe the difference in refraction for air, fresh water and salt water and relate those observations to practical applications.

MATERIALS:
§ soup bowl
§ penny
§ saltwater solution (50 grams of table salt to 1 liter of fresh water)

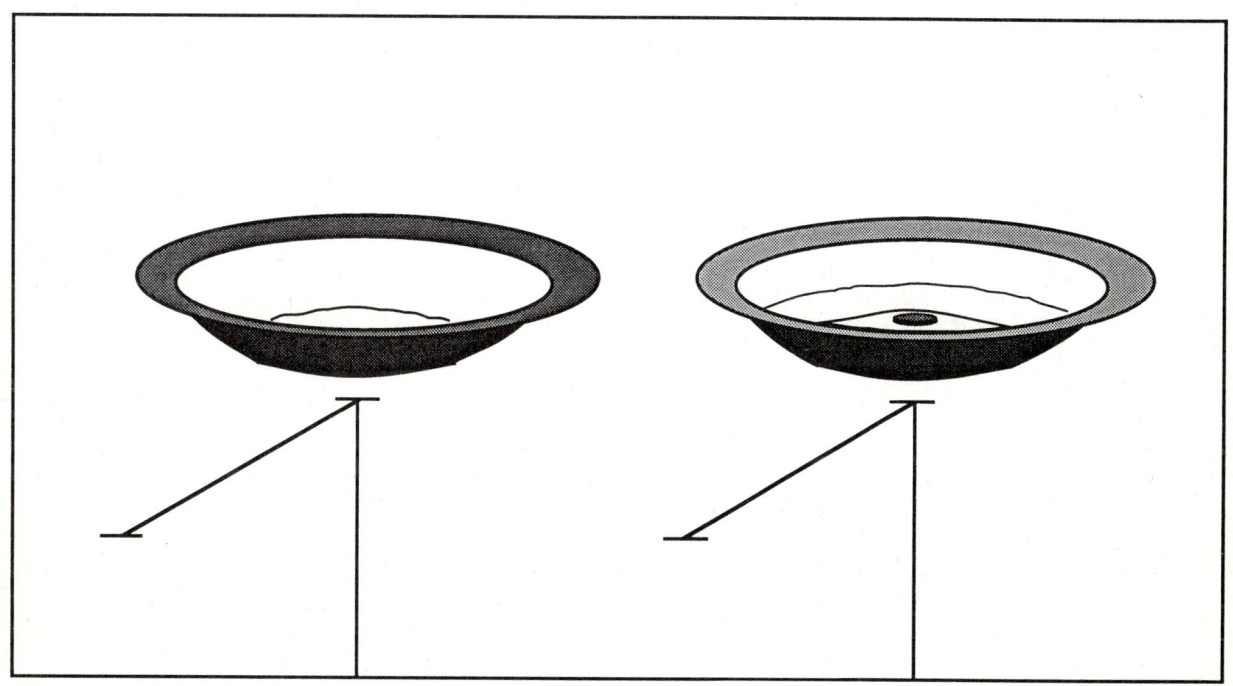

Fig. 5.7

## PROCEDURE:
1.  Place a penny in center of a bowl (Fig. 5.7).  Hold the bowl straight out from your chest at arm's length.  The penny should not be visible.
2.  Mark a spot on the wall or blackboard to indicate the level to which you raised the bowl. Return the bowl to this height for each procedure.

3. Add 1/4 cup of water to the bowl, making sure the penny stays in the center. Again, hold the bowl at arm's length. Do you see the penny? If not, begin adding as many tablespoons of water as necessary to make the penny visible. How many additional tablespoons did it require?

4. Repeat the procedure using salt water. How many tablespoons of salt water did it require?

QUESTIONS:

1. Which medium -- air, fresh water, or salt water -- has the highest index of refraction? Explain.

2. In this experiment, light rays reflect off the penny and travel toward the observer's eyes. Draw the light rays. (Remember: the light ray bends as it breaks the surface of the water, not when it reaches the top of the bowl.)

3. A tropical fish collector was rowing along the shore when he spotted a rare fish he wanted. The fish was swimming 8 feet below the surface of the water. Given that the angle of refraction (as the light travels from the fish to the collector's eyes) is 40 degrees, draw a ray diagram. Show the real and apparent positions of the fish.

To calculate the apparent position, first use Snell's Law to determine the angle of incidence. Then use the given angle of refraction (40 degrees) and the angle of incidence you calculated above. Hint: You will need to use basic trigonometry to solve this problem, i.e., relationships between angles.

# LAB 19: AEROSOLS, VISIBILITY AND THE COLOR OF THE SKY

## INTRODUTION

The color of the sky and visibility vary greatly with the time of day, the presence or absence of clouds, and particulate matter or aerosol in the air. In the absence of clouds the sky can appear red, blue, green, or even gray. These changes in color are related to refraction and scattering of light waves, which also affect visibility and exposure to burning, ultraviolet radiation. During the day a clean, clear atmosphere appears blue due to the preferential scattering of blue light by air molecules. The wavelength of blue light is similar to the diameter of air molecules making air molecules an efficient scatterer of blue light. As the sun sets, the sun's rays pass through more and more air, diminishing the blue light through scattering, until red wavelength light dominates. Aerosols add to the scattering. If the aerosol is small, such as those that result from natural turpines released into the air by forests, the result is a blue cast to the air. The Blue Mountains of North Carolina and the Blue Mountains of Australia gain their name from this phenomena.

When particles in the air are large relative to the wavelengths of visible light, they scatter all these wavelengths equally; thus cloud droplets result in white clouds (unless they are tall, in which case the bottoms appear dark) during the day. Haze droplets make the sky appear gray through scattering. Human activity (burning of fossil fuels, agricultural activity, etc.) has increased the burden of aerosol in the atmosphere during this century, resulting in a decrease in the amount of sunlight reaching the Earth's surface. Recent research has shown that areas where the aerosol input to the atmosphere is greatest have experienced cooler surface temperatures. This effect may partially offset global warming.

## ACTIVITY

OBJECTIVE: The purpose of this activity is to observe the effects of scattering and absorption on light waves, using water as an analogy for air.

MATERIALS:
§ 2 to 3 clear glass containers
§ bright (Halogen) flashlight

§ toilet paper roll
§ cardboard
§ 2 to 3 pipettes or dyedroppers
§ milk
§ mud

## PROCEDURE:

1. To intensify the light source, make a collimator. Take the toilet paper roll and cut two pieces of cardboard into circles for end pieces. Cut a vertical slit 1/8 inch wide in each of the circular ends. Glue the ends to the toilet paper roll, aligning the vertical slits.
2. Fill the glass containers with water. Shine the light through the containers. (Note: Place the collimator between the flashlight and the glass container. It will focus the light into a more concentrated beam.) Because tapwater contains little particulate matter, not much light will be scattered or absorbed.
3. Add one to two drops of milk to the container and stir well. Shine the light through the container (Fig. 5.8).

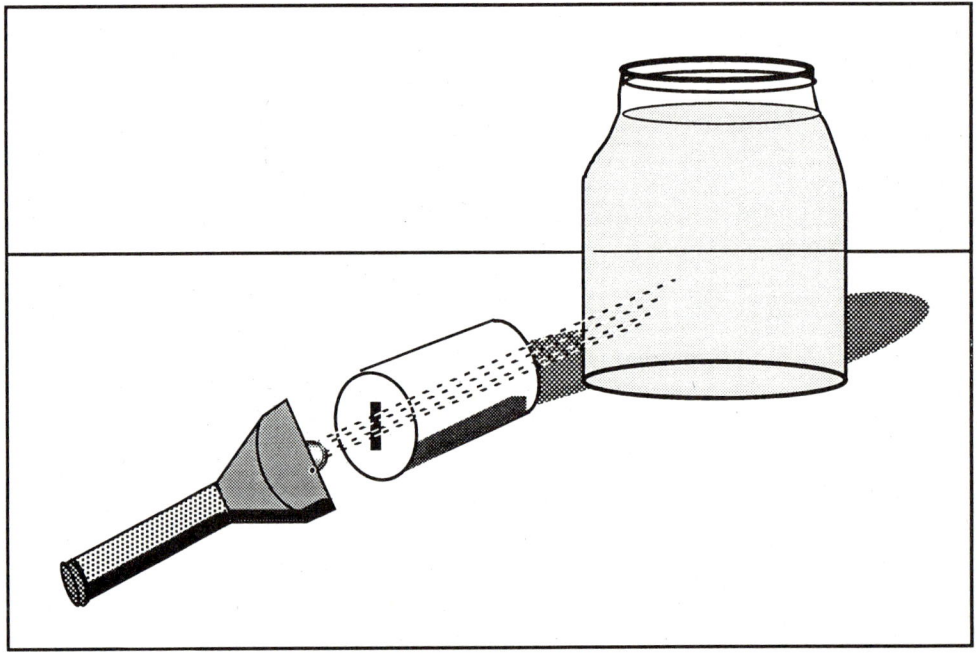

Fig. 5.8

Look at the water in the container from the top or from a side perpendicular to the beam of light. *What color do you see?*

Look at the water in the container from the side opposite the flashlight. *What color do you see?*

Place additional glass containers behind the first to accentuate the color.

Repeat the above procedure using mud.

QUESTIONS:

1. What is the color of the milky water when you observe it from the top or side? Explain

2. What is the color of the milky water when you observe looking through the water toward the flashlight? Explain

3. After adding mud to the glass, what is the color of water from the top, looking down and at the sides? Explain

4. Given the statement that smaller particles scatter blue light and larger particles scatter all colors, try adding other impurities to clean water and see if you can make statements about the size of the impurities added based on your observations of the resulting colors.

# *Lab 20:  Observing Weather Phenomena*

## INTRODUCTION

One of the primary goals of this manual is to promote an awareness of the natural environment around us.  Since the atmosphere is continually changing, it is possible to walk outside at any moment and observe some type of meteorological or atmospheric optical phenomena.  Not all phenomenon are as spectacular as a rainbow or a snowstorm; however, variations in the shades of the sky or the evolution of a cloud provide interesting subjects.  Although it may require some patience to observe some of the phenomena that occur in our environment, once you've learned to look for them you may well be surprised at how commonly some optics, such as halos, appear.

## ACTIVITY

OBJECTIVE:  The purpose of this lab is to promote an awareness of our natural environment.

MATERIALS:
§ pen & paper
§ camera or miscellaneous artistic supplies
§ creativity

Make a point of personally observing an atmospheric, geophysical, or atmospheric optical phenomena.  Any observation that you personally witness during the current semester, and has a relation to the subject matter of this course, is fair game.  Creativity in the choice of your observation and in doing the assignment will be considered in grading.  Some examples include these:

Atmospheric optics--rainbows, halos, color variations in the sky, etc.
Cloud formation and motions, condensation of water on a glass
Dispersion of pollution--cigarettes, smokestacks
Behavior of waves breaking, stream flow, effects of wind on water
Climatic effects on land forms
Doppler effects--passing cars, trains, etc.

Remember, these are just a few suggestions and you are definitely not limited to the above list.

## PROCEDURE:

1. Describe the setting of your observation, including the date, time, place and any circumstances which contribute to the phenomenon. (*Note: this observation must be made during the current semester, however, it need not necessarily take place outdoors*)

2. Describe the phenomenon in detail. Pay careful attention to detail. Use of your creative writing abilities is encouraged.

3. Illustrate (sketch, photo, etc.) the phenomenon or a relevant aspect of the setting or physical mechanism(s) involved. Try to preserve the relative scales in the sketches. Creativity in any artwork will be appreciated. Photography is encouraged.

4. Give a concise explanation of what physically is producing the phenomenon. The more specific your choice of observation, the easier it will be to fulfill each of the four steps. The final page should be about one type written page, plus an illustration.

### The water (hyrdrological) cycle

The total quantity of water in the Earth ecosystem is assumed to be unchanging. Nonetheless, there is a continuous exchange of water from the oceans and land to the atmosphere as a result of evaporation and transpiration. The water returns to the surface layers in the form of rain and snow. The sum of the water exchanges over the Earth is called the *hydrologic cycle*. On the average over the continents, precipitation exceeds evaporation. The excess water runs off into the oceans to compensate for the excess of evaporation over precipitation over the oceans.

For the Earth as a whole, average yearly precipitation amounts to a depth of water equal to 100 cm (0.27 cm/day). On the average, if all the water vapor in a column of air from the ground to the top of the atmosphere were condensed, it would produce a column of water 3 cm deep. If the atmospheric water vapor were rained out at the rate of 0.27 cm/day, it would take 11 days to "empty" the atmosphere of water vapor. This period is known as the average turnover or residence time of water vapor in the atmosphere. It is a statistically derived quantity which shows that water molecules, once evaporated into the air, tend to stay there a fairly short time and reflects the dynamic nature of the hydrological cycle. Even so through the action of winds, water molecules are likely to leave the atmosphere as rain or snow at a point far from where they entered it.

On the average over the continents, precipitation exceeds evaporation. The excess water runs off into the oceans to compensate for the excess of evaporation over precipitation over the oceans.

## Clouds

Clouds and precipitation are a dynamic part of the hydrological or water cycle on Earth. A cloud is composed of a very large number of small water droplets or ice crystals. The size and shape of the cloud depends on the characteristics of the atmosphere, particularly the moisture content, and the pattern of vertical air motions.

There are various ways to classify clouds. The most widely used one groups clouds according to their appearances. There are 10 principal cloud types whose names were derived from Latin in the 1800's by an English scientist named Luke Howard *(cirrus, cirrocumulus, cirrostratus, altocumulus, altostratus, nimbostratus, stratocumulus, stratus, cumulus, and cumulonimbus)* that can be regarded as falling in three major cloud subdivisions:

i) *cirrus* (hair-like) - high feathery clouds composed of ice crystals and usually occurring above ~6 km.

ii) *stratus* (layered) - clouds that form in layers

iii) *cumulus* (pile or heap) - a detached cloud having the appearance of a mound, dome, or tower.

Some clouds have the characteristics of more than one of these categories. When a cloud is producing rain or snow, the term *nimbus* is added. For example, *nimbostratus* is a layer cloud producing precipitation. The prefix alto is added when clouds occur in the range from ~2 to 6 km in elevation.

After sunset and before sunrise in polar regions, beautiful clouds sometimes are seen at great altitudes, as they are brilliantly illuminated by the sun against the night sky. *Nacreous* clouds occur in the lower stratosphere at altitudes of 20 to 30 km, while *noctilucent* (literally "night light") clouds appear at altitudes close to 80 km.

Small cumulus clouds over continents are composed of water droplets mostly smaller than 20 $\mu$m (micrometers) in diameter that are present in concentrations of more than 1000 per cubic centimeter. In cumuli over tropical oceans, droplet concentrations may be as small as 50 per cm$^3$, with the largest cloud droplets having diameters exceeding 50 $\mu$m. These differences are related to the greater abundance of particles in the air over continents than over oceans.

Tiny particles in the atmosphere, called *condensation nuclei,* are preferred sites for condensation to occur because they help reduce the retarding impact of surface tension on cloud droplet growth. They are introduced into the air from the ground and the oceans, as well as by combustion, both natural and manmade. Tiny particles may also be formed in the atmosphere from sulfur dioxide and nitrogen dioxide gases. Those particles, salty and acidic, on which condensation occurs readily are called hygroscopic nuclei. Condensation begins on such nuclei when the relative humidity of the air approaches saturation. Seasalt left in the atmosphere when ocean spray evaporates is an important source of hygroscopic condensation nuclei.

As water droplets are carried above the altitude where the temperature is 0°C, they generally do not freeze immediately. Instead, they remain liquid even at subfreezing temperatures and are said to be *supercooled.* Clouds often are supercooled to temperatures of -10°C to -15°C; on rare occasions, super cooling extends to temperatures as low as -40°C.

Ice crystals in the atmosphere occur at subfreezing temperatures above -40°C when *ice nuclei* are present. Ice nuclei are particles whose crystalline shape mimics that of ice. Certain soil particles are effective natural ice nuclei. Substances such as silver iodide, when they are dispensed as finely divided particles, can produce ice crystals at temperatures as high as -5°C. For this reason, they can be used to modify supercooled clouds.

Ice crystals develop in a number of shapes, each of which displays distinctive hexagonal (six-sided) characteristics. The most common shapes are the following: columnar—a long, narrow prism of hexagonal cross section; plate—a thin, solid plate having six sides; dendritic—a six-sided star, sometimes with each arm having an intricate, lacy structure.

### Precipitation

Precipitation in the form of rain or snow occurs when particles of water or ice are large enough to reach the ground. Wisps or streaks of falling precipitation particles that evaporate before reaching the ground are called *virga*. The chief physical difference between a cloud element and a precipitation element is size. In terms of water drops, the boundary between cloud droplets and raindrops is generally assumed to be a diameter of ~200 $\mu$m. Water drops having diameters from 200 $\mu$m to 500 $\mu$m are sometimes called drizzle drops.

It takes many cloud droplets to make up a single raindrop. A raindrop 2 mm in diameter contains a million times more water than does a droplet having a diameter of 20 $\mu$m. In order for rain to occur, other drop growth processes, in addition to condensation, must occur.

Raindrops can be produced by the *collision* and *coalescence* of cloud droplets. Collisions take place because the terminal velocity of a water drop increases as its diameter increases, over the normal size range of cloud droplets and raindrops. Large droplets fall faster than, collide with and merge with smaller ones. When two rain droplets merge, coalescence has taken place. As a result of coalescence, the large drops can grow fairly rapidly.

Raindrops are also produced by the melting of ice crystals, snowflakes, and other frozen particles. When ice crystals exist in subfreezing air in the presence of supercooled water, the crystals grow as the droplets evaporate. This occurs because, at the same temperature, *the saturation vapor pressure over ice is less than that over water.* As a result, there is a pressure force driving the water molecules from the water to the ice, resulting in a rapid growth of ice crystals in the presence of liquid cloud droplets.

As ice crystals grow, the heavier ones fall. As a result, collisions and accretions occur. A snowflake is an aggregate of ice crystals stuck together. When such a particle falls through a layer of air whose temperature is above freezing, the crystals melt and raindrops are produced. In mountainous areas during the winter, valley locations often experience rain while snow falls at higher elevations.

Sometimes, in association with warm fronts in winter storms (see section 7), snowflakes descending through a warm layer aloft melt and become raindrops. If they fall into subfreezing air under the front, the raindrops may be supercooled and become *freezing rain.* The water freezes when it falls on cold surfaces and produces hard, solid ice that is particularly hazardous to pedestrians and motor vehicles.

Occasionally supercooled raindrops freeze in the air and reach the ground as small pellets of ice. Such precipitation is technically called *sleet,* but this word is sometimes used to mean a mixture of rain and snow.

Hail particles are ice pellets greater than 5 mm in diameter. Along the eastern slope of the Rockies in the vicinity of the Wyoming and Colorado borders, where hail is relatively frequent, hailstones exceeding 2 cm in diameter are fairly common. Such storms occur over most of the United States on occasion. In exceptional instances, hailstones may be as large as oranges. Hail occurs with disturbingly high frequency in many other countries. It does substantial damage to crops in northern Italy and Switzerland, Argentina, Kenya, South Africa, and elsewhere.

Hail is formed in thunderstorms containing strong, persistent updrafts. In such situations, hailstones can be sustained in supercooled environments long enough to allow them to "sweep out" a great deal of supercooled water. The large hailstones, e.g., those greater than 2 cm in diameter, are believed to form most often in storms having updrafts tilted into the wind. This would allow some hailstones to make several passages through an updraft before falling to the ground.

# Lab 21: The Water Cycle

## INTRODUCTION

The water (hydrologic) cycle is the movement and exchange of water substance among the atmosphere, ocean, and lands. Water vapor in the atmosphere plays a major role in the water cycle by returning water from land and ocean reservoirs across the globe, to even the highest elevations. Water vapor in the air condenses as it cools to form clouds, which return rain and snow to the Earth's surface.

Invisible to the eye, water vapor enters the atmosphere by evaporation. Evidence of this includes the often observed drying of sidewalks and roads after rain. Not so easily detected is the continual evaporation of water from soils, directly and through plants (transpiration). Once in the air, water vapor is carried aloft into the atmosphere through updrafts (or convection) and to other locations by the wind currents (advection). Ultimately, it changes phase to liquid or ice when the air is cooled to its dew point temperature. The following activity shows these processes on a small scale.

## ACTIVITY

### OBJECTIVES:

The purpose of this activity is to obtain evidence that water vapor enters the atmosphere from a variety of land surfaces and to describe, based on observations, differences in evaporation rates from various surfaces.

### MATERIALS:

§ Clear plastic sheeting material
§ Black plastic sheeting material
§ Small weights

### PROCEDURE:

Cut three one-meter square pieces of sheeting material. Place one over a grassy surface, one over bare soil, and one on an asphalt or concrete sidewalk. Anchor the corners with weights to keep the sheets in place. After one half hour, observe the sheets of plastic and compare the relative amounts of condensation, if any, on their lower surfaces.

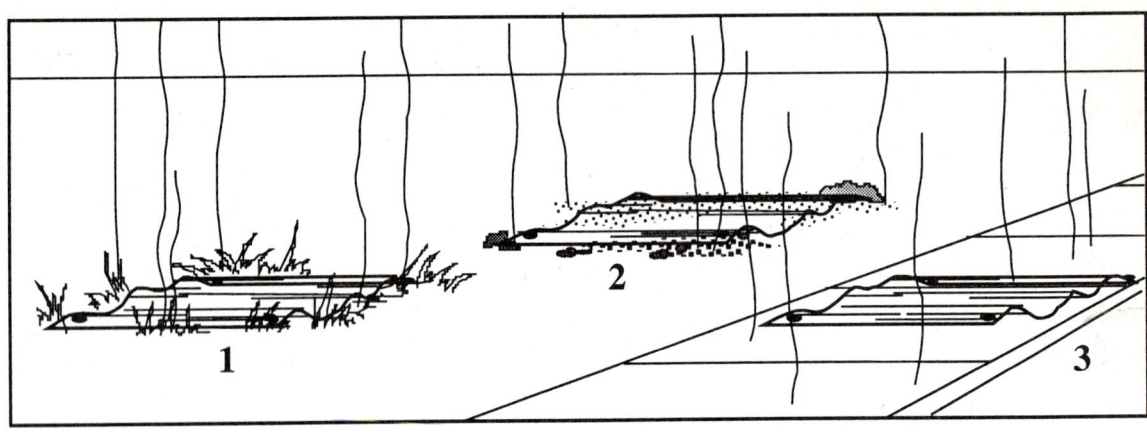

Fig. 4.5

QUESTIONS:

1. Which land surface, if any, produced the most liquid water on the underside of the plastic sheets? The least?

2. Where did the water, if any, come from? What phase changes did it have to go through to end up deposited on the plastic sheets?

3. What can you conclude about different kinds of land surfaces and their relative influences in transferring water to the atmosphere?

4. How can your observations and conclusions be applied to describe the natural water (hydrologic) cycle?

5. Try this investigation with both clear and black plastic sheets. Do they produce different effects on the amount of evaporation? If so, why?

Extra Credit:
6. Try this investigation at various times of the day or under various cloud cover conditions. How does the brightness of sunshine appear to affect the rate of evaporation from the surfaces? Why?

7. Construct a broad, shallow "evaporation pan" and fill with water. Measure the drop in water level from day to day, and determine the rates of evaporation from the water surface under different weather conditions.

8. Design and conduct a version of this investigation to actually measure the rates of evaporation from different kinds of surfaces. One way might be to collect and weigh the water which accumulates in an hour above each surface.

# Lab 22: Formation of Clouds

## INTRODUCTION

There are three conditions in the atmosphere that are met before a cloud forms. First, there must be sufficient moisture in the air. Secondly, the air must cool so that it becomes saturated and condensation can occur. And finally, there must be some type of particulate or condensation nuclei suspended in the air such as dust, smoke, or pollen for the excess water to condense on.

## ACTIVITY

OBJECTIVE: The purpose of this experiment is to investigate the conditions which must be present in order for clouds to form.

MATERIALS:
§ 32-oz. clear glass jar with lid
§ ice cubes or crushed ice
§ hot water
§ matches
§ can of aerosol spray (air freshener is best)
§ black construction paper
§ safety goggles
§ flashlight (optional)

PROCEDURE:
1. Fill the jar with hot water.
2. Pour out the hot water leaving only ~2 cm of water in the bottom of the jar. Place the jar in front of the upright construction paper.
3. Turn the lid of the jar upside down and fill it with ice. Place the lid on the jar as shown in Fig. 6.1. Observe the jar for three minutes. Darken the room and shine the flashlight on the jar as you make your observations. Record your observations in the Data Table.
4. Pour the water out of the jar and repeat steps 1 and 2.
5. Move all loose papers away from the jar. Wearing your safety goggles, strike a match and drop the burning match into the jar.

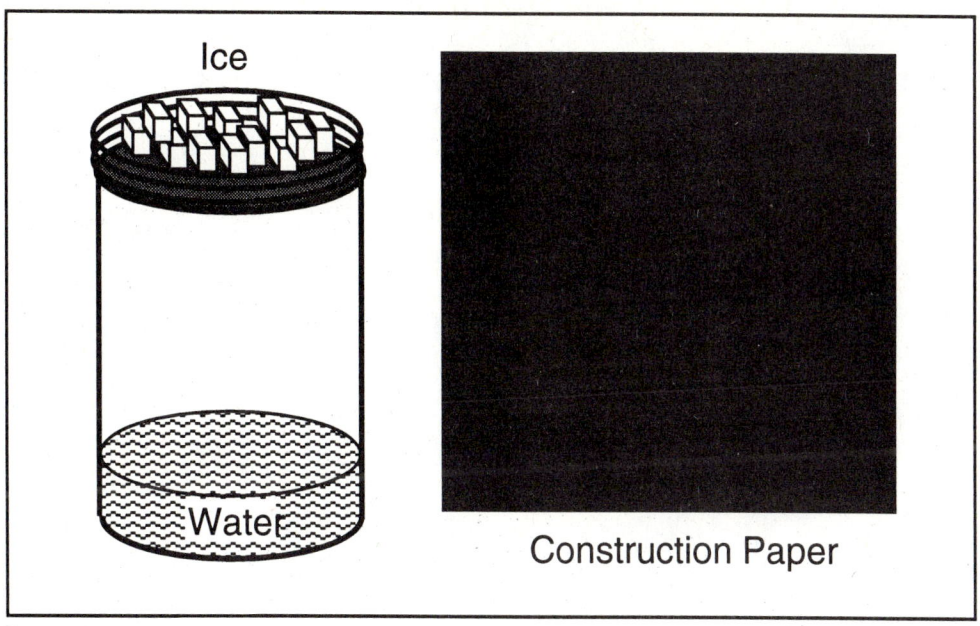

Figure 6.1

6. Immediately cover the mouth of the jar with the lid full of ice as you did in step 3 and observe what happens for three minutes. Record your observations in the Data Table.
7. Pour out the water and repeat steps 1 and 2.
8. Spray a very small amount of aerosol in the jar and immediately cover the mouth of the jar with the ice filled lid. Observe what happens for three minutes and record the observations in the Data Table.

**Data Table**

| TRIAL | OBSERVATIONS |
|---|---|
| No match or aerosol | |
| Match | |
| Aerosol | |

QUESTIONS:

1. Comment on the differences on your observations, and the reasons for these differences.

2. Suppose a layer of sand was put on the bottom of the jar instead of water, would a cloud form? Why or why not?

3. What would have happened in this experiment if cold water had been used instead of hot water? Why?

4. Describe any motion observed inside the jar. Explain your observation.

# *Lab 23: Investigating Hail*

## INTRODUCTION

As mentioned previously, hail particles are ice pellets that are greater than five millimeters in diameter. Hail forms in thunderstorm cells where there are strong updrafts, great vertical development, and an abundant supply of supercooled water droplets (T<32°F). The updrafts allow the hail particle to grow while it is suspended by the updraft. Therefore the stronger the updraft is, the larger the particle will be.

Hail causes millions of dollars of damage each year in the United States alone. Eighty percent of the damage from hail destroys crops and farm land. In the midwest, farmers buy hail insurance and liability coverage in order to avoid a total loss of income as a result of a hailstorm.

Hail pellets range in size from pea size to golf ball size to softball size. The largest hailstone on record fell in Coffeyville, Kansas on September 3, 1970. It was six inches in diameter and weighed 1.67 ponds. In June, 1959, the town of Selden, Kansas was covered in hail nineteen inches deep. On June 13, 1984, golf ball size hail fell on Denver, Colorado for an hour and a half. This storm accumulated ten inches of hail, three foot drifts, and approximately 350 million dollars in damage. Although hail is dangerous, in the United States only two deaths have been reported as a result of hail. In China, however, a violent hailstorm in June of 1932 killed 200 people.

Despite the prevalence of hail, it requires very specific conditions in order to form. First, there must be *supercooled* water droplets in the atmosphere. In the absence of a freezing nuclei to initiate crystallization, a clean liquid water droplet surrounded air will not freeze even when it is cooled well below 0° C. Such liquid droplets below freezing are referred to as supercooled. Second, there must be something in the atmosphere on which the water may crystallize. Usually this requirement is met by special particles that mimic the shape of ice crystals. Finally, there must be strong updrafts in the atmosphere. If a strong updraft exists, it can catch the small piece of ice and carry it back into the cloud. Many additional supercooled cloud droplets will then crystallize on it, adding a new layer of ice each time it rises and falls again. This cycle is responsible for the layered growth of hailstones. The cycle continues and the hailstone grows in size until the force of gravity overcomes the force of the updraft.

**ACTIVITY**

OBJECTIVE: The purpose of this experiment is to investigate some of the interesting and essential factors in the production of hail. The same principles demonstrated here also apply to the formation of most rain over the United States, in that 95% of our rainfall begins in the upper parts of clouds as snow

MATERIALS:
§ crushed ice
§ large test tube
§ 400-600 ml beaker
§ salt
§ distilled water
§ thermometer
§ stirring rod

PROCEDURE:
1. Fill the beaker with equal amounts of water and ice.
2. Pour in enough salt so that after stirring, you can still see salt on the bottom of the beaker. It is better to have too much salt than not enough.
3. Wash the test tube, being sure that no dust or dirt remains on the inside. Fill the test tube with cold distilled water so that the level of water in the test tube is the same as the level of water in the beaker when the test tube is placed in the beaker (Fig. 6.2).
4. Put the thermometer in the beaker and then put the test tube in the beaker as shown in Fig. 6.2.
5. Allow this to sit for ~ six minutes, stirring occasionally with the stirring rod. The temperature in the beaker should fall below zero.
6. At the end of six minutes, remove the thermometer and record the temperature in the Data Table.
7. Remove the test tube and immediately drop a small piece of crushed ice into it. Record your observations in the Data Table.
8. If time allows, empty the test tube and repeat steps 3 through 7.

Figure 6.2

**Data Table**

| Trial | Temperature (˚C) | Observations |
|-------|------------------|--------------|
| 1 | | |
| 2 | | |

QUESTIONS:

1. What was the temperature in the beaker at the end of six minutes? What physical principle caused the temperature to drop below zero?

2. What happened when the piece of ice was placed in the test tube?

3. The water in the test tube was below freezing point before the piece of ice was inserted. Why do you think that the water did not freeze before the ice was inserted?

4. Why do you think it was important to clean the test tube so well before you used it for this activity? Why did you use distilled water?

5. In this experiment, "hail" formed in the test tube. How is this different from what happens in the atmosphere?

# Lab 24: Recycled Water-The Hydrologic Cycle

## INTRODUCTION

In considering the hydrology of the Earth, it is relevant to recognize that the oceans cover about 70% of the Earth's surface and that they represent more than 97% of the Earth's water. Another 2.2% is locked up in ice caps and glaciers. As global temperatures increase or decrease, the quantity of land-held ice decreases (by melting) or increases (by precipitation), resulting in important changes of sea level. The atmosphere contains only 0.001% of the planet's water, but this water crucially affects the lives of people, animals, and plants.

In the activity which accompanies this demonstration, a distillation apparatus is used to model the hydrologic cycle. It is important to keep in mind the distinctions between distillation and the hydrologic cycle as it occurs in nature.

## ACTIVITY

OBJECTIVE: The objective of this activity is to investigate the hydrologic cycle.

MATERIALS:
§ clear plastic shoe box with lid
§ small plastic cup
§ Baggie filled with sand or soil
§ water
§ ice
§ lamp with reflector

PROCEDURE:
1. Set up the apparatus shown in Fig. 4.3.
2. Cut a hole in one corner of the lid of the shoe box, just large enough for the cup to fit halfway through the lid.
3. Add enough water to cover the bottom of the clear shoe box, a depth of about one inch.

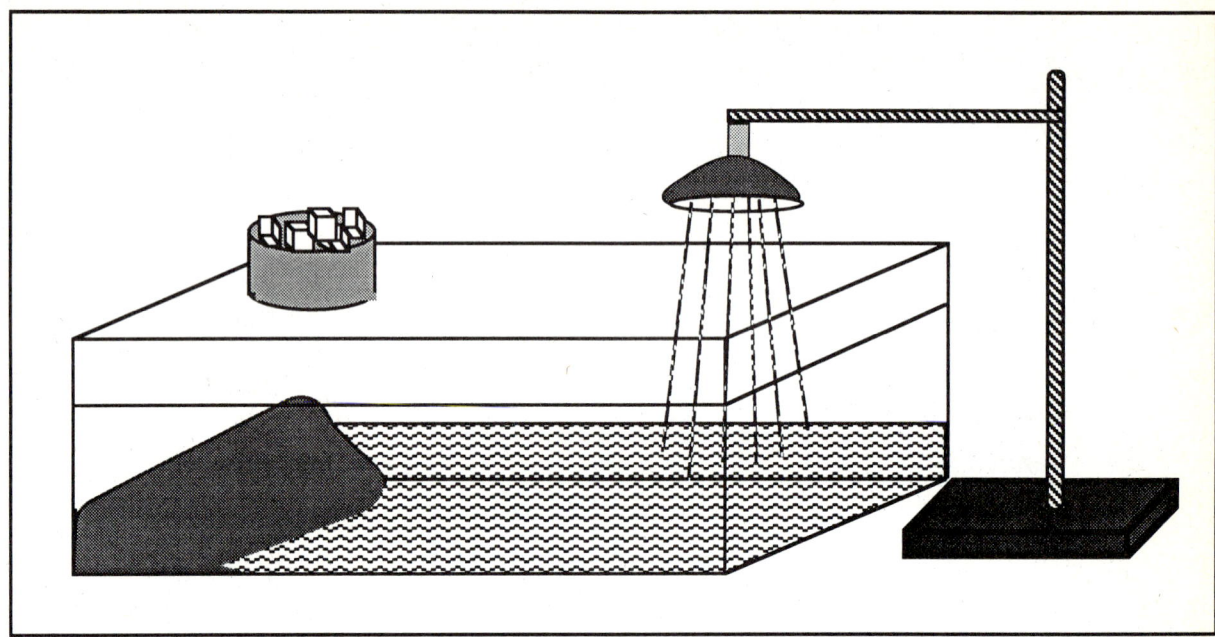

Figure 6.3

4. Position the sandbag at one end of the box, directly under the opening of the cup.
5. Fill the small plastic cup with ice and place it in the opening of the lid.
6. Position a gooseneck lamp so that its light is shining onto the water inside the box.
7. Periodically check the set-up to observe the progression of events. Record your observation over the course of the class and diagram the movement of the water through the set-up.

QUESTIONS:
1. What causes water droplets to form on the outside of the cup which is in the shoe box?


2. Where in this activity does evaporation occur? Condensation?

3. Where in the Earth's hydrologic cycle does evaporation occur? Condensation?

4. From what you observed in this activity, what are the key processes in the hydrologic cycle?

5. Was water lost from the system in any part of this activity? If so, where?

The term wind is generally used to describe the horizontal motion of air. Vertical air motions usually are called updrafts or downdrafts. Wind velocity has both a direction and a magnitude, therefore, it is a *vector* quantity. Wind speed is measured with an instrument called an *anemometer*. Most often, winds in the lower atmosphere have speeds of a few meters per second (m/s), but in the upper atmosphere they often exceed 50 m/s (112 mph), and in tornadoes and in hurricanes they may exceed 100 m/s (224 mph). Wind direction is measured with a *wind vane* and is given as the direction *from* which the wind is blowing.

In the eighteenth century, Sir Isaac Newton discovered three laws of motion:

  I)  In the absence of forces, an object at rest will remain at rest and an object in motion will remain in motion with the same velocity.

II)     The acceleration of a body is equal to the force acting on it divided by its mass.

III)    To every action there's an equal and opposite reaction.

Based on these laws, the velocity of the air usually can be calculated if one knows the forces acting on it. The five forces governing winds are these:

1)  *pressure-gradient force* - occurs because atmospheric pressure varies from place to place.

2)  *gravity* - only acts in the vertical direction and is closely balanced by the vertical gradient in pressure. This balance of forces is called hydrostatic balance.

3)  *Coriolis force* - exists because the Earth rotates under a moving air stream, resulting in an apparent deflection of the wind as observed from the ground. In fact, the Coriolis force is not a true force in the sense of the pressure-gradient force or gravity. It is sometimes called the Coriolis effect. If the wind were measured with respect to a point fixed in space there would be no Coriolis force. In such a circumstance, an air parcel with no forces acting on it would move in a straight line at a constant speed. But we measure wind velocity with respect to the surface of the Earth. From this reference frame, the effect of the Earth's rotation is equivalent to a force. The following observations summarize the effects of the Coriolis force: a) It is proportional to the speed of the moving object, b) is zero at the equator and increases to a maximum at the poles, and c) always acts at 90° to the motion, resulting in a deflection of air parcels to the right of their motion in the Northern Hemisphere and to the left in the Southern Hemisphere. In the absence of other force, air parcels would display a curved path to the right in the Northern Hemisphere.

4)  *centrifugal force* - is an apparent force that is invoked when the wind follows a curved path such as those generally found in cyclonic storms. Newton's first law of motion states that a parcel in motion will continue in a straight line unless acted upon by an unbalanced force. In the vicinity of a low pressure center in the atmosphere, the orientation of the pressure gradient force changes as the flow moves around the low. This provides the unbalanced

force that causes the wind to follow a curved path. The centrifugal force is a fictitious force that acts opposite this changing pressure gradient force and actually reflects the air's inertia or tendency to move in a straight line.

5) *frictional force* - is present in all moving systems and acts to oppose the motion. In the atmosphere, friction is most important near the surface (lowest ~1 kilometer of the atmosphere) and causes dissipation of the energy of motion into heat energy.

When isobars are straight and frictional forces are negligibly small, the wind velocity is determined only by the pressure-gradient and Coriolis forces. In such cases, the pressure gradient and Coriolis forces balance each other, and the resultant wind is said to be geostrophic. The geostrophic wind is parallel to the isobars with low pressure on the left if you stand with your back to the wind in the Northern Hemisphere. The closer the isobars are to one another, the higher the wind speed. In the Southern Hemisphere, the same result occurs, except that low pressure is on the right when the wind is at your back.

When isobars are curved, such as they are around a circular region of low pressure, the effects of curvature must be taken into account. When the air is moving along a curved path, the net force must be producing an acceleration towards the center of rotation. When the wind is blowing parallel to curved isobars under the action of the pressure gradient, centrifugal and Coriolis forces, it is called the *gradient wind*. The wind is nearly gradient in character at altitudes above one kilometer, where frictional effects of the ground are small.

Frictional forces act in a direction opposite to the direction of the wind-velocity vector. They act to reduce the velocity, which, in turn, reduces the Coriolis force. As a result, the wind is deviated across the isobars towards low pressure. The frictional effects are a maximum near the ground and decrease with height. In the Northern Hemisphere, when the directions of the isobars do not change with height, the wind vector turns clockwise with height up to about one kilometer, where the winds are in gradient or geostrophic balance. The rotation of the wind vector near the ground sometimes is called the *Ekman spiral.*

Along coastlines, wind velocities often deviate markedly from the geostrophic or gradient values. Differential heating of land and water can produce air temperature and pressure differences, and a convection cell is established. Cool air from over the water moves over the land in the form of a *sea breeze.* Aloft, air moves over the sea and sinks, providing a good example of a convection cell. Similarly, at night the land cools faster than the water surface, and a *land breeze* is established. In a sense, monsoons are sea and land breezes on a massive scale. In the winter, cool, dry continental air tends to flow towards warmer oceans; in the summer, warm, humid oceanic air blows over the land.

In hilly and mountainous regions, special winds arise as air sinks or rises along the slopes. The *Foehn, chinook, and Santa Ana* winds flow downhill in response to the larger-scale circulation and are warm and dry. In desert areas where radiation effects cause pronounced warming and cooling, mountain and valley winds are common. During the day, winds blow up the sun-heated valleys; at night, cool, heavy air sinks down the valleys. Special winds also result when cold, dense air originating over elevated (often over ice) plateaus descends to lower elevations. Such winds are called *katabatic* winds and are pervasive along the coasts of Greenland and Antarctica.

### General wind circulation of the Earth's atmosphere

The atmosphere can be regarded as a heat engine which receives energy from the sun and converts it into the kinetic energy of air motions. Of the vast energy received on the average from the sun, only two to three percent of it can account for the kinetic energy represented by the wind systems over the Earth.

As noted in Section 3, at high latitudes the Earth radiates more energy to space than it absorbs, while at low latitudes it absorbs more than it emits. As a result, there must be heat transported poleward in order to prevent long-period temperature changes. The heat transport is accomplished mostly by ocean and air currents. If the Earth were a uniform stationary object, you would expect a huge convection cell with air rising at low latitudes, moving poleward aloft, sinking at high latitudes, and returning equatorward near the ground. You would also expect a convection cell between the day side and night side of the Earth. But the Earth is a non-uniform, rotating spheroid, and as a result the air motions are not in the form of simple convection cells.

Interestingly, there is a convection cell, called the *Hadley cell*, at low latitudes. Air rises over equatorial regions, moves poleward aloft, sinks at latitudes of about 30°, and flows equatorward near the ground in the form of the trade winds (see Fig. 7.1). In the Northern Hemisphere, the trades blow mostly from the northeast; in the Southern Hemisphere, from the southeast. The trade winds converge along the *intertropical convergence zone* (ITCZ) or equatorial trough, where winds are light and the air is warm, humid, and uncomfortable. This region is sometimes called the *doldrums*.

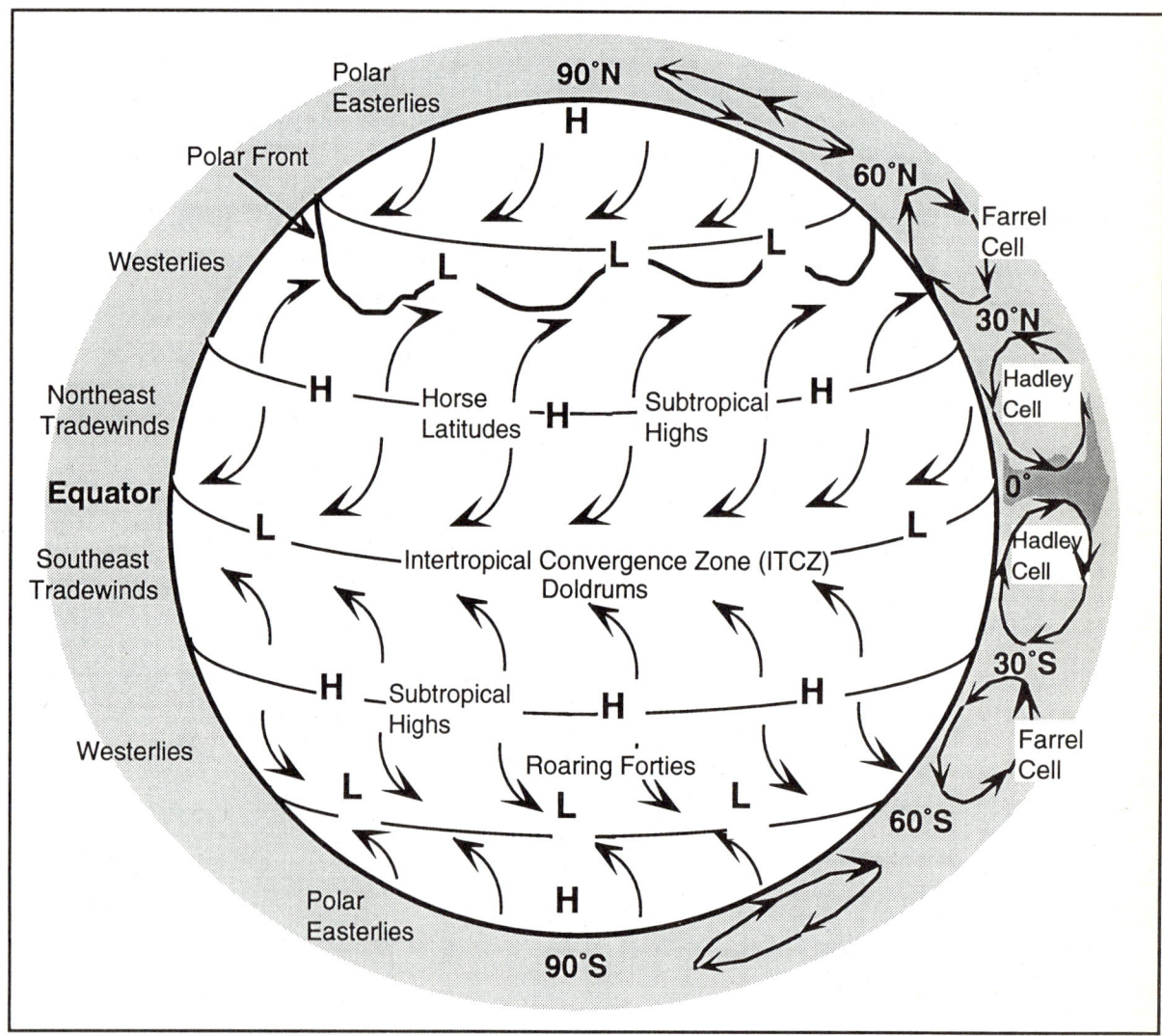

Figure 7.1 General circulation of the atmosphere

In the latitude belt near 30°, where sinking air predominates, large, semi-permanent anticyclones prevail. Within them, the air tends to be dry and the winds light over very large areas. These areas were named the *horse latitudes* by the sailors of long ago.

Poleward of the Hadley cell, there is a broad latitude region dominated by westerly winds that increase with height and have embedded within them traveling cyclones and anticyclones. On the average, in the belt of westerlies, the winds near the ground have a component away from the equator and a reverse component aloft. Poleward of about the 60° latitude circle, there is a third cell with easterly winds predominating near the surface.

Observations of pressure and wind patterns reveal significant variations from season to season. The wind fields follow the sun. For example, the ITCZ has an average position at about 5° S latitude in January and moves northward as the months go by, reaching a position averaging about 10°N latitude in July.

The strength of the winds, as noted earlier, depends mostly on the pressure-gradient force. In the winter, the pressure-gradient force between latitudes of the westerly winds (about 30° to 60°) is at a maximum. This occurs because the north-south temperature gradient is a maximum. As a result, the westerly winds are strongest during that season. The highest velocities tend to be concentrated in a restricted current called the *polar-front jet stream,* found at altitudes of about 12 km, over the portion of the polar front across which north-south temperature changes are concentrated. The polar-front jet stream exhibits average maximum winds of about 60 m/sec, but can be much stronger. It tends to follow a meandering path around the globe at middle latitudes. A second strong westerly wind current, the *subtroptropical jet stream* is found at lower latitudes and slightly higher altitude than the polar-front jet.

Jet streams, particularly the one over the polar front, have important effects on the global weather patterns by contributing to the transport of atmospheric properties and the formation of storm centers. Jet streams have both good and bad influences on aviation. By planning flights to maximize tail winds, airplane ground speeds can be optimized in order to reduce flight times. On the other hand, in regions where the wind-velocity gradient is large, that is, where the *wind shear* is large, there can be strong *clear-air turbulence* (CAT). Pilots seek to increase ground speed while avoiding turbulence.

### Air masses, fronts and midlatitude cyclones

The state of the atmosphere can be depicted on synoptic maps, that is, maps showing patterns of various weather elements at a particular time. Surface weather stations and rawinsonde stations all over the world follow observing practices formulated by the World Meteorological Organization (WMO) in Geneva, Switzerland. The observations are transmitted via radio and computer to central receiving points such as Washington, D.C., Moscow, Russia, and Melbourne, Australia, and then on to many other stations. The data are plotted on weather charts according to standard WMO procedures. Meteorologists analyze the maps by drawing such features as isobars, isotherms, the location of fronts, and centers of high and low pressure.

An examination of a weather map shows that there are large regions over which temperature and humidity change relatively little. For example, the southeastern United States, in summer, is often covered by a warm, humid mass of air moving in from the Gulf of Mexico and Caribbean Sea. At the same time, over the northeastern United States and Canada, there could be a large region of cool, dry air. The term *air mass* is used to represent these distinctive bodies of air.

An air mass develops its principal characteristics by remaining over a particular region of the Earth long enough for the air to approach temperature and moisture equilibrium with the underlying surface. This occurs as a result of heat transfer by various mechanisms (described in Section 2) and by evaporation or condensation.

Air masses are classified mostly according to the region of formation and whether or not the underlying surface is oceanic (maritime) or continental. Four principal air masses and the letters used to designate them are:

*i) maritime tropical—mT*      *iii) maritime polar—mP*

*ii) continental tropical—cT*      *iv) continental polar—cP*

Two additional air mass categories sometimes referred to are *arctic* and *equatorial* corresponding to air masses that form over polar ice sheets at high latitude, and over equatorial rain forests, respectively. Once an air mass leaves its source region, it can be modified by exchanges of heat and moisture with the surface over which it passes. For example when a Canadian continental-polar air mass passes over the warm waters of the Gulf Stream it is rapidly modified by the transfer of heat and moisture to the air from below.

It is found that when air masses having significantly different properties encounter one another, they do not mix readily. Instead, the cooler air wedges itself under the warmer air, and the two air masses are separated by a zone of transition between them that is called a *front*.

There are various types of fronts, whose names depend mostly on the direction of movement:

*Cold front*—the cold air advances, replacing warm air.

*Warm front*—the warm air advances, replacing cold air.

*Stationary front*—the transition zone between cold, and warm air remains essentially stationary.

*Occluded front*—the frontal system resulting when a cold front overtakes a warm front, forcing the warm air aloft.

Frontal zones are responsible for the formation of much of the cloudiness, rain, and snow that occur over the United States, especially in winter. The warm, humid, less dense air is forced to rise over the cold air. In the process, condensation and precipitation take place.

Frontal zones also are important because they are favored locations for the formation of *cyclones*. A cyclone is a region of low pressure around which there is closed circulation. In the central United States, the term cyclone is sometimes used to mean a tornado; in southeast Asia, the name cyclone refers to tropical storms in the hurricane family. When a meteorologist speaks of cyclones, he usually is talking about nearly circular regions of low pressure a few to several hundred kilometers in diameter. In middle latitudes, cyclones often form along pre-existing frontal systems and are the source of a great deal of clouds, rain, and snow.

Cyclones develop along frontal zones because denser, cold air is located at the same height as nearby, less-dense, warm air. Such a condition represents a source of potential energy. The heavy air sinks, displacing the warm air, and converts potential energy into kinetic energy in the form of a cyclonic wind circulation. The following sequence of events is often seen on weather maps (Fig. 7.2).

A small wave develops on a stationary front. At the surface, pressures begin to fall. A continuation of this process for a day or two leads to the formation of a wave cyclone. As the wave grows, it takes on the appearance of a breaking wave with a cold front advancing and overtaking a warm front .

Over the Earth as a whole, frontal cyclones occur most frequently in regions where major frontal systems tend to stagnate. For example, in the winter, polar air from North America pushes far to the south. The associated front tends to stagnate along the eastern slope of the Rockies and through the Gulf of Mexico. Cyclone initiation is common over Colorado, Alberta, Canada, in the Gulf of Mexico, and off the east coast of the United States.

The polar front extends across the Pacific and Atlantic Oceans and defines a preferred belt of cyclonic development. Many cyclones that eventually affect North America originate in the north Pacific, in the vicinity of the Aleutian Islands. They account for a semi-permanent center of low pressure called the Aleutian low. It has an Atlantic counterpart known as the Icelandic low.

Cyclones, particularly those of the frontal variety, are dynamic in character. They change rapidly under the influence of various energy sources and sinks. Some have brief lifetimes lasting for less than a day and producing no "weather." Others may last for more than a week, growing in size and intensity as the central pressures fall. Such storms may become greater than 1000 km in diameter and produce heavy rain, snow, and strong winds.
A special group of cyclones are called thermal lows. They occur in summer over hot regions such as the deserts of Arizona and northwestern Mexico. The vertical extent of a thermal low is very shallow and its circulation weakens rapidly with height. Often times a thermal low becomes an anticyclone when its circulation reverses direction.

Severe tropical cyclones such as hurricanes and typhoons make up another class of cyclones. They do not form along frontal zones. They develop over warm oceans and derive most of their energy from the water below. This will be discussed in more detail in section 8.

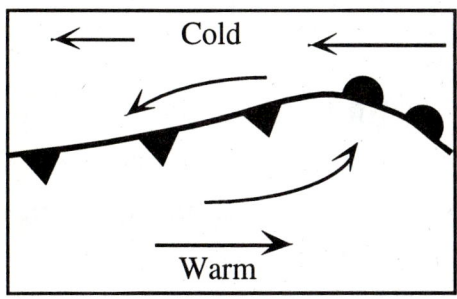

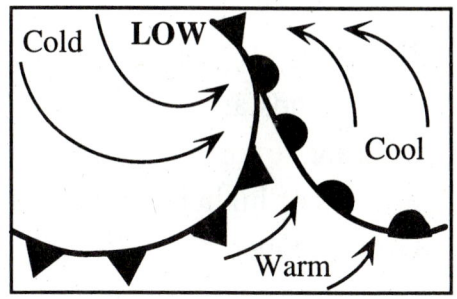

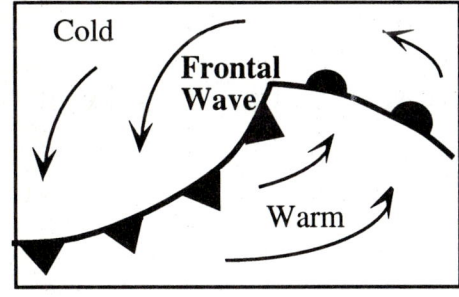

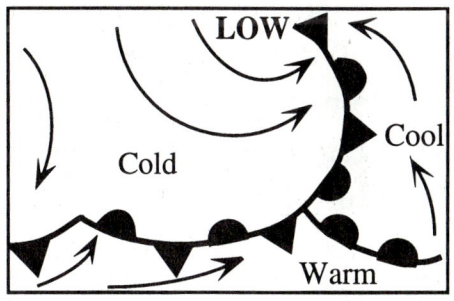

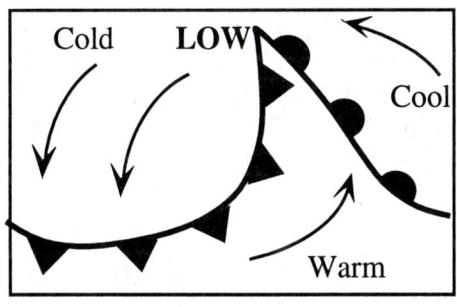

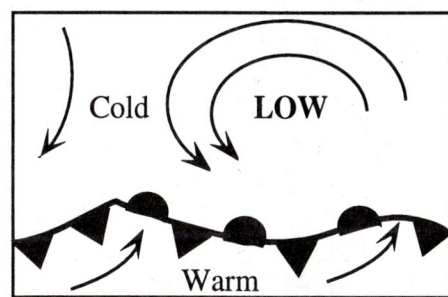

Figure 7.2  Schematic of the evolution of a midlatitude cyclone or winter storm

111

# Lab 25: Wind: Air in Motion

## INTRODUCTION

Winds arise from pressure gradients. As areas of high and low pressure migrate, evolving pressure gradients between them will force the air in regions of higher pressure to move toward regions of lower pressure, creating winds. Once the air is moving, other forces come into play. The Coriolis force, due to the rotation of the Earth under moving air, prevents the air from taking a path directly toward lower pressure, thus contributing to the spiral wind patterns seen in storm systems.

When a balloon is blown up, the pressure of the air in the balloon becomes greater than the pressure of the air surrounding the balloon. Therefore, when the balloon is opened, the high pressure air rushes out into the surrounding region of lower pressure producing a puff of wind.

There are, however, differences between the model of wind in this activity and winds in the atmosphere. The material boundary of the balloon that gives rise to the pressure gradients in this activity has no counterpart in the atmosphere. Thus, the creation of wind is usually not as abrupt in the atmosphere as when the balloon being released. The acceleration of wind occurs gradually as a storm develops and its central pressure drops.

## ACTIVITY

OBJECTIVE: The objective of this activity is to investigate the processes involved in creating wind.

MATERIALS:
§ balloon (Long ones work best, as opposed to round ones)
§ string or fishing line (5 meters)
§ drinking straw
§ clear tape

PROCEDURE:
1. Thread the string through the straw and have one person hold one end and you hold the other.

2. Blow the balloon up, but do not tie it. With the help of your partner, tape the balloon to the straw (Fig. 7.3).
3. Pull the string tight and move the straw to one end of the string.
4. Let go of the balloon and observe what happens. **Record your observations.**
5. Repeat the process two more times. **Record any additional observations.**

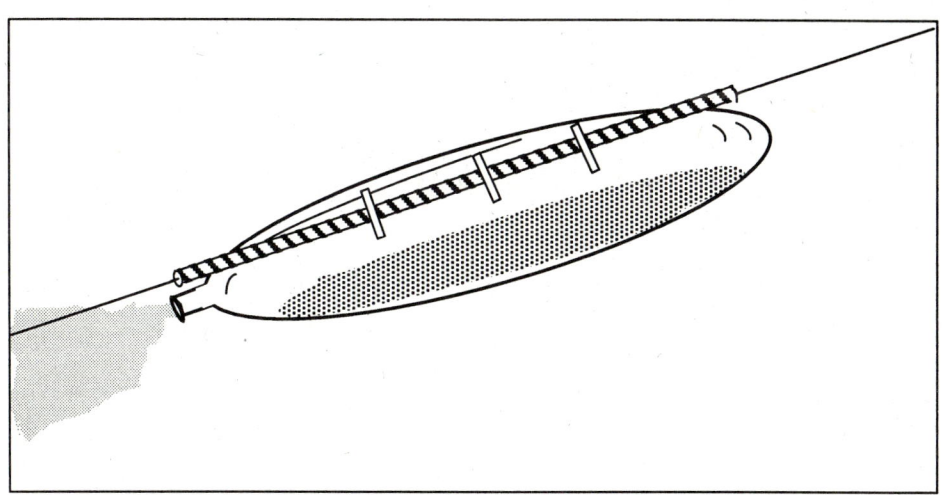

Fig. 7.3

QUESTIONS:

1. What happened to the air in the balloon when the balloon was released? Why?

2. In your trials, what force(s) caused your balloon to slow or stop?

3. What forces that are generally important for winds in the atmosphere are not important here?

# Lab 26: The Coriolis Effect

## INTRODUCTION

As winds and ocean currents move across the Earth's surface, they cover long distances over long periods of time and experience a deflection due to the rotation of the Earth underneath the air. The farther from the equator the motions occur, the greater the deflection is. The deflection causes winds flow in a counterclockwise direction around low pressure centers in the Northern Hemisphere and in a clockwise direction in the Southern Hemisphere.

Imagine that you and a friend are standing inside a large box that is mounted on a carousel. If the carousel is slowly rotating, you can imagine not being aware of it, just as you are unaware that the Earth is rotating daily on its axis. If you and your friend begin throwing a softball back and forth, it will appear to be deflected to one side, as if pushed off course by some unseen force, making it hard to catch.

A similar unseen force acts on moving air currents in the atmosphere. It is called the "Coriolis" force and the deflection it causes is the Coriolis effect. Like centrifugal force, the Coriolis force is only an apparent force and the Coriolis effect is only an apparent deflection. Yet both are important in understanding the motion of air parcels in the atmosphere.

Although the Earth is rotating, we do not take the effects of this rotation into account in our daily lives because it is relatively slow. For example, when playing baseball we do not notice a deflection of the ball due to the rotation of the Earth even though it does occur. That is because a baseball travels across the field quickly, not allowing the Earth to rotate very far underneath the ball during this short period. Since winds and ocean currents travel over long periods and distances the effect becomes important and can be seen in the structure of atmospheric storm systems and in the ocean circulation.

## ACTIVITY

OBJECTIVE: The purpose of this activtiy is to investigate how objects are deflected in a rotating frame of reference.

## MATERIALS:

§ obtain or construct a Coriolis demonstrator that has a freely rotating base. A "Lazy Susan" or a round swivel stool will work.

§ attach a plate or "magic slate" to the rotating base. The plate ideally should be about 18-24" in diameter and have a surface that can be marked (Fig. 7.4).

§ a ball bearing (inch) or a marble

§ chalk or other writing implement to mark the surface of the plate

## PROCEDURE:

Roll a ball bearing or marble across the surface of the plate with no rotation. Spin the top of the rotating plate slowly. Draw a chalkline directly across the rotating plate. Even though you moved the chalk in a straight line, the mark was curved (S-shaped). Experiment with spinning the plate in different directions. Or start your mark from the middle of the rotating plate.

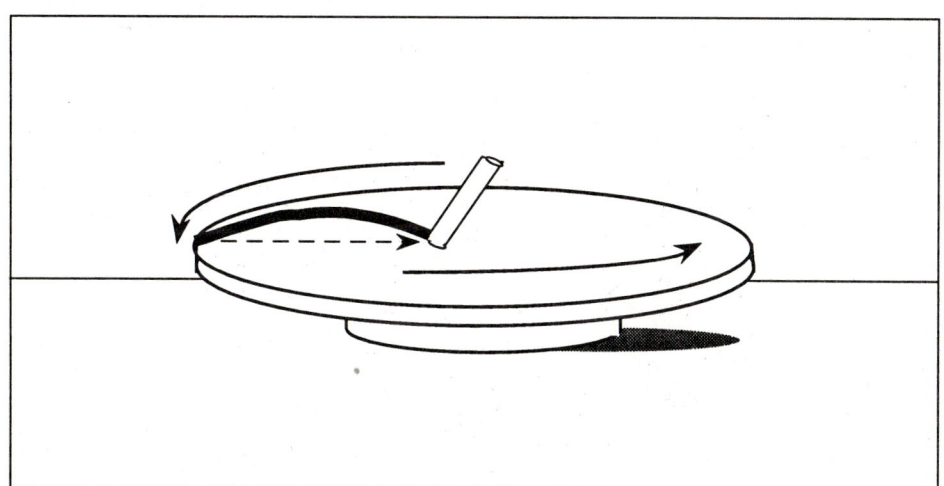

Fig. 7.4

## QUESTIONS:

1. Describe the path of the ball bearing in the case of no rotation?



Rotate or spin the plate in a counterclockwise direction and roll the ball bearing across it.

2. Describe the ball bearing's motion.

3. Predict the motion of the ball bearing if the plate was rotated in a clockwise direction.

   Rotate the demonstrator in a clockwise direction.
4. How does your prediction compare with what happened?

5. The center of the plate represents the poles of the Earth. Predict what would happen if you released the ball bearing from the center of the plate when it was rotating clockwise. Predict what would happen if you released the ball bearing from the center of the board when it was rotating counterclockwise.

6. What happened?

   Release the ball bearing from the center of the plate when it is rotating counterclockwise.
7. What happened?

8. When you turn the demonstrator in a counterclockwise direction, which hemisphere is it representing? And if you turn it clockwise, which hemisphere is represented?

# Lab 27:  Analyzing a Winter Storm

**INTRODUCTION**

There are many types of weather charts or maps that the National Weather Service distributes to aid operational forecasters.  One of the most important types is an *analysis* of the current weather showing observations collected by the global network of instruments.  An analysis can depict many things related to the weather, including frontal positions and contours of constant pressure and temperature, etc.  Another type of weather chart shows numerical prognoses or forecasts of  the future state of the weather through a period of several days, and is generated by numerical weather predictions models run on super computers at the National Meteorological Center near Washington D.C.  In this lab, we will concentrate on a surface analysis chart.

On the surface analysis chart observations from around the country are plotted, allowing the meteorologist to see weather patterns across the United States.  Once the observations are plotted on the map, meteorologists analyze the map to determine areas of low and high pressure, temperature, dew point, etc.  Contours are drawn to show lines of constant pressure, called *isobars* and lines of constant temperature, called *isotherms*.

Data on a surface analysis are plotted in a uniform manner around the point locating the observing station on the map.  Below is an example of such a *station model* and an explanation of the plotted data.

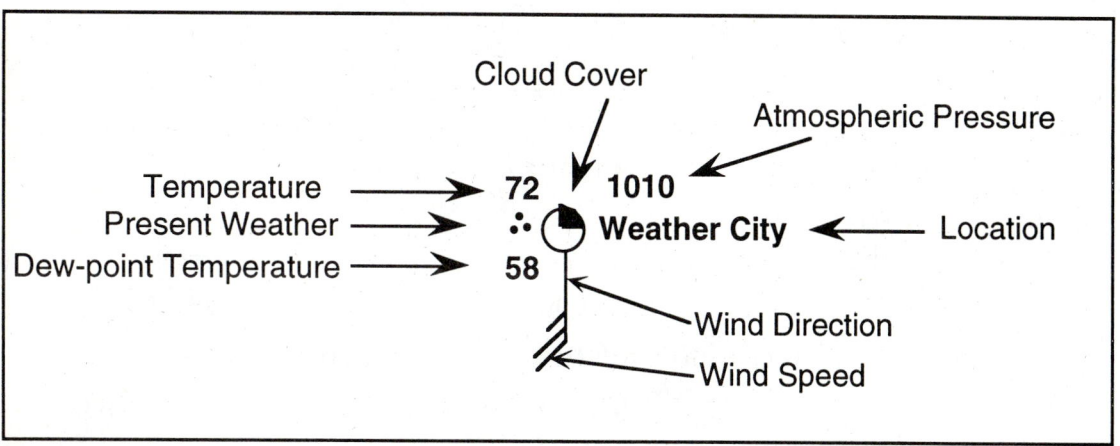

Figure 7.5  Station Model for plotting of surface weather observations

*Cloud Cover* - This represents the amount of cloud cover over the station. One quarter of the circle is filled in for one quarter cloud cover, one half for one half cloud cover, etc. If the sky is obscured an X is put in the circle.

*Atmospheric Pressure* - This is the atmospheric pressure measured in millibars (mb). Meteorologists draw contours of equal pressure called *isobars*. In practice the number recorded on the station model includes only the last two numbers and the decimal place. However, for this lab, the entire pressure is plotted in whole millibars.

*Temperature* - This is the temperature measured in °F at the top of each hour. Isotherms designate areas of equal temperature.

*Dew-point Temperature* - This is the dew point temperature measured in °F at the top of the hour. You will recall that the dew point temperature is the temperature at which water droplets form.

*Wind Direction* - This line represents the direction from which the wind is blowing.

*Wind Speed* - The small barbs represent the wind speed. Each full line represents ten knots (1 kt =1.15 mph). Shorter lines represent wind speed increments of 5 knots. If the winds speed exceeds 50 knots, a triangle shaped barb is used. The total wind speed is determined by adding the barbs (See Fig. 7.6).

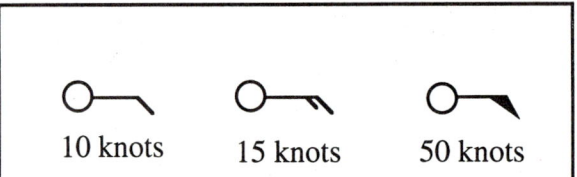

Figure 7.6

If you think of the wind speed barbs as feathers on an arrow, the circle represents the arrowhead. The arrow points the direction the wind is blowing to. In meteorology the wind direction is designated as the direction from which the wind is blowing. Therefore, if an arrow points to the *west*, the wind direction is actually called *east* (Fig. 7.6).

*Present Weather* - Symbols are used to show the weather that is occurring at the time of observation (Fig. 7.7).

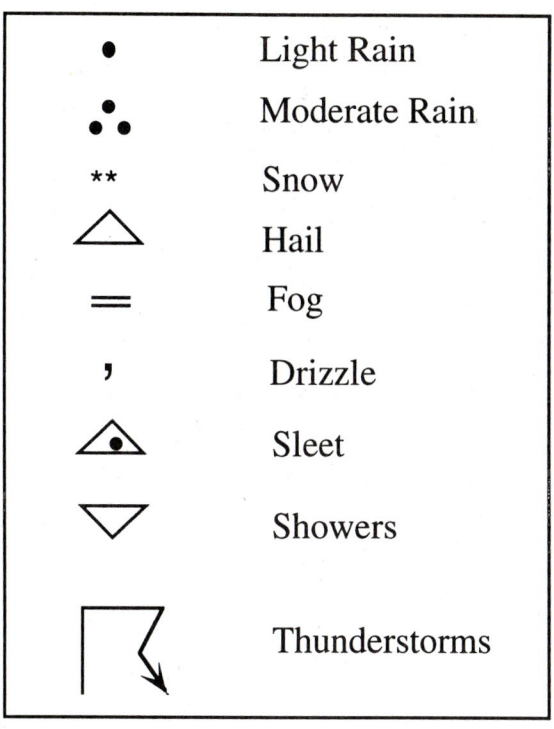

Figure 7.7

Charts that depict the current conditions in the atmosphere above the surface are also very valuable to the forecaster, and are similar to surface maps in that they show observations plotted for a particular time. Surface maps are issued every three hours, whereas upper-air maps are issued only twice daily corresponding to the times at which radiosonde balloons are launched. In addition to data collected by radiosonde instruments, upper air maps also show data from aircraft reports and satellites.

## ACTIVITY

<u>OBJECTIVE:</u> The purpose of this activity is to learn how to interpret and analyze data plotted on a surface chart and see the evolution of a winter cyclone and its fronts.

<u>MATERIALS:</u>
§ #2.5 or #3 pencil (harder lead will make analysis easier to erase),
§ #2 pencil (to finalize analysis)
§ red and blue pencils, and large pink or white erasure

119

<u>PROCEDURE:</u>

Plotted data are provided for four times during the evolution of a rapidly deepening winter storm that brought heavy snow and severe weather across the central United States.

*1. Analysis Background*

a. There is no general rule about the correct place to start. However, it is usually best to begin at a place where the pressure distribution seems apparent and the reports are ample. Use a harder lead pencil and draw lightly so that it is easy to make erasures.

b. Lines of constant pressure on a surface map are called isobars. To analyze the pressure field, think of pressure as you would altitude on a contour map of a smooth mountain. In areas where tight pressure gradients exist, a good practice is to draw the isobars for larger intervals and then draw in the intermediate values. Fig. 7.8 provides a sample analyzed weather map for 15 December at midnight (00) Greenwich Mean Time (GMT)* (7 AM EST), which correspondes to the plotted map in Fig. 7.9a.

c. Isobars are smooth, continuous lines except where they cross sharply defined fronts and at the edge of the map. They should never cross or join other isobars.

e. Surface winds will cross the isobars from high to low pressure at an angle of ~20°-30°. The size of the angle depends on the roughness of the terrain; the rougher the terrain the greater the angle between the isobar and the wind direction.

f. The speed of the wind is inversely proportional to the spacing of the isobars.

g. Plotted values may be in error due to the method by which sea-level pressure values are obtained. The error may be particularly large in mountainous sections of the country. In all cases, consider the possibility of error by the observer in reading, coding, decoding, or plotting the pressure values.

---

* Greenwich Mean Time (GMT) is the time standard used by meteorologists to deal with the problem of multiple time zones when collecting and analyzing weather data across large distances.

h. The occurrence of thunderstorms at or near the station may cause pressures which may vary considerably from surroundings reports.

i. Lines of constant temperature are called isotherms. Isotherms can include more wiggles than isobars, reflecting the impact of local surface variations on this variable.


2. *Sea-Level Pressure Analysis*

a. Analyze the surface-pressure fields in Figs. 7.9a-d. Draw isobars every 8 mb (begin with an isobar value divisible evenly by 8, e.g., 1000, 1008, 1016...). Analyze quickly just to get the sense of the distribution.

b. Now go back and add isobars for every 4 mb, keeping in mind that you wish to show as smooth a pattern as possible without undue disregard of the data. Strive for gradual changes in the spacing between isobars.

c. Label the high and low pressure centers. Use block type letters to designate location of highs (blue H) and lows (red L).

d. Label all contours. Label open isobars at both ends; they may also be labeled at other points. Label closed isobars at a convenient point that does not overwrite plotted data. Break the isobar at that point to permit the entry of the value. For a series of closed isobars, the labels should be arranged for form an easily read line of numbers (see Fig. 7.8).


3. *Surface Temperature Analysis*

a. Analyze the temperature field in Figs. 7.9a-d. The standard contour interval for isotherms is 5° F. Therefore, draw contours every 5° F (e.g., ...40, 45, 50, 55, 60°, .....etc.). Label all contours as before.


4. *Finding the Cold Front*

Using wind shifts, temperature contrasts and current weather as clues, locate and place the cold front in dark blue pencil on your analyzed maps (study Fig. 7.8 as an example). Use the previous frontal position as a guide to future positions; fronts tend to move in one direction.

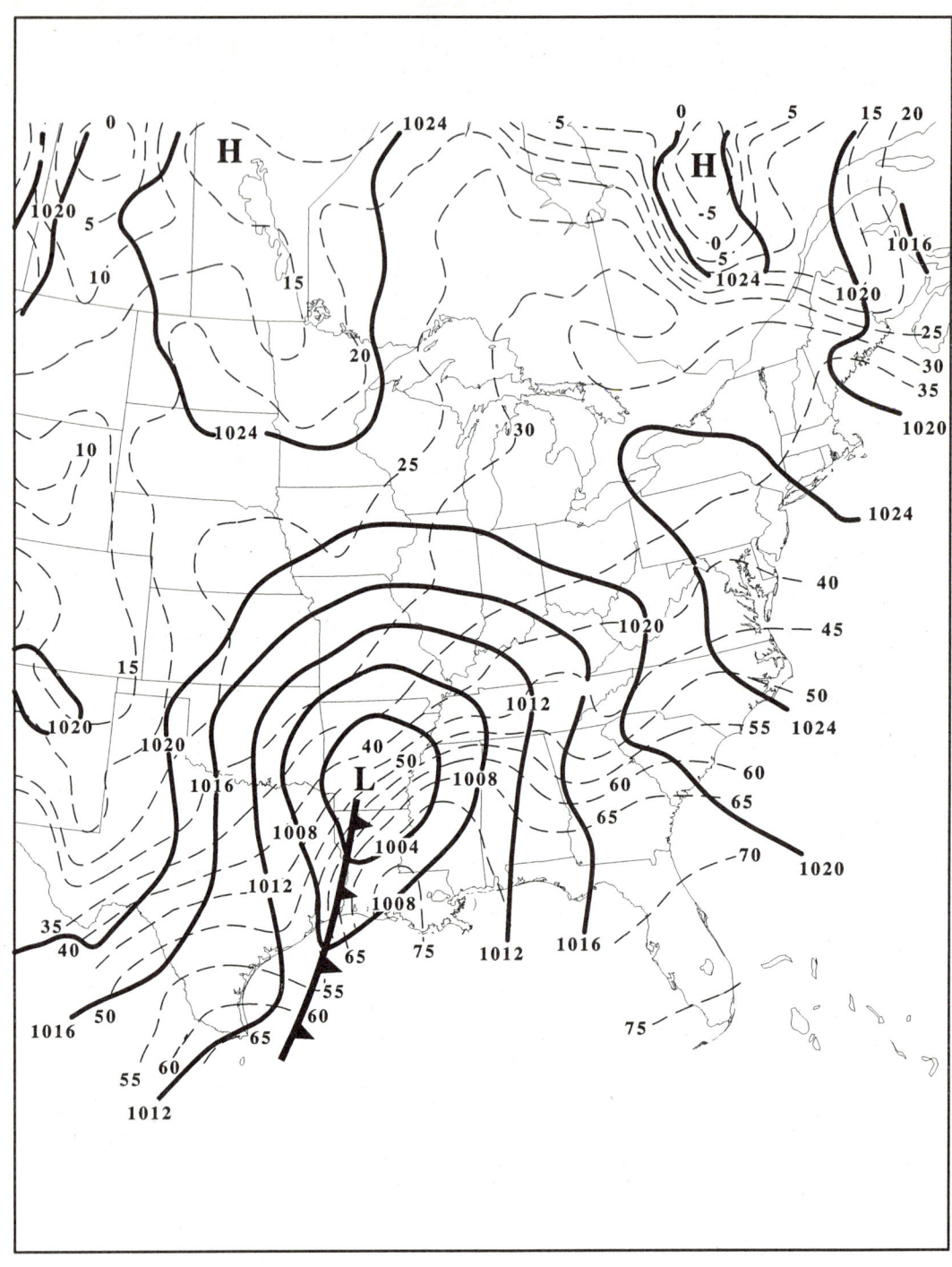

Fig. 7.8 Analysis of sea level pressure (solid contours every 4 millibars) and temperature (dashed contours every 5° F) for weather observations at 00 GMT (7 AM EST) 15 December 1987.

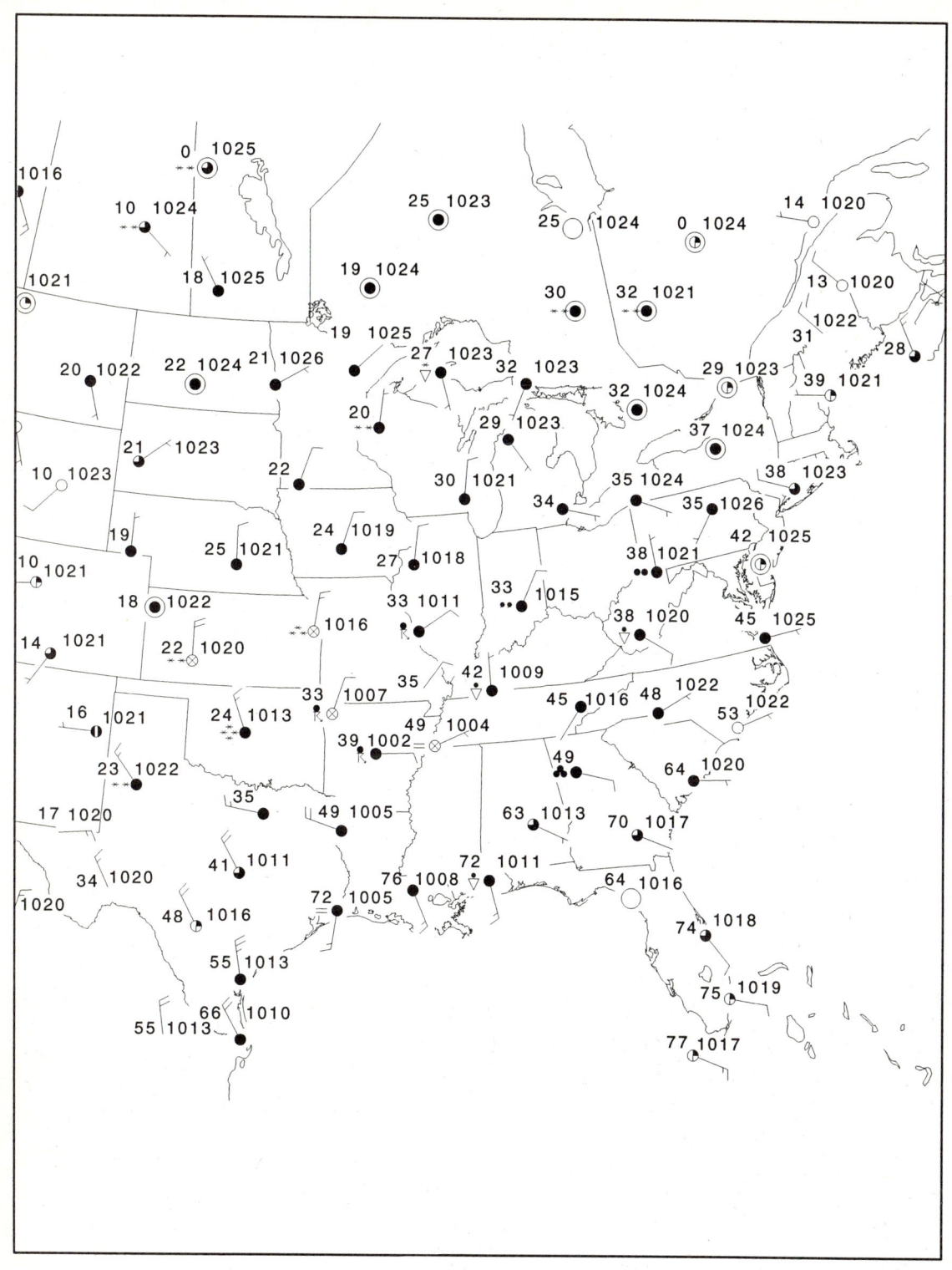

Fig. 7.9a  Plotted weather observations for 00 GMT 15 December 1987.

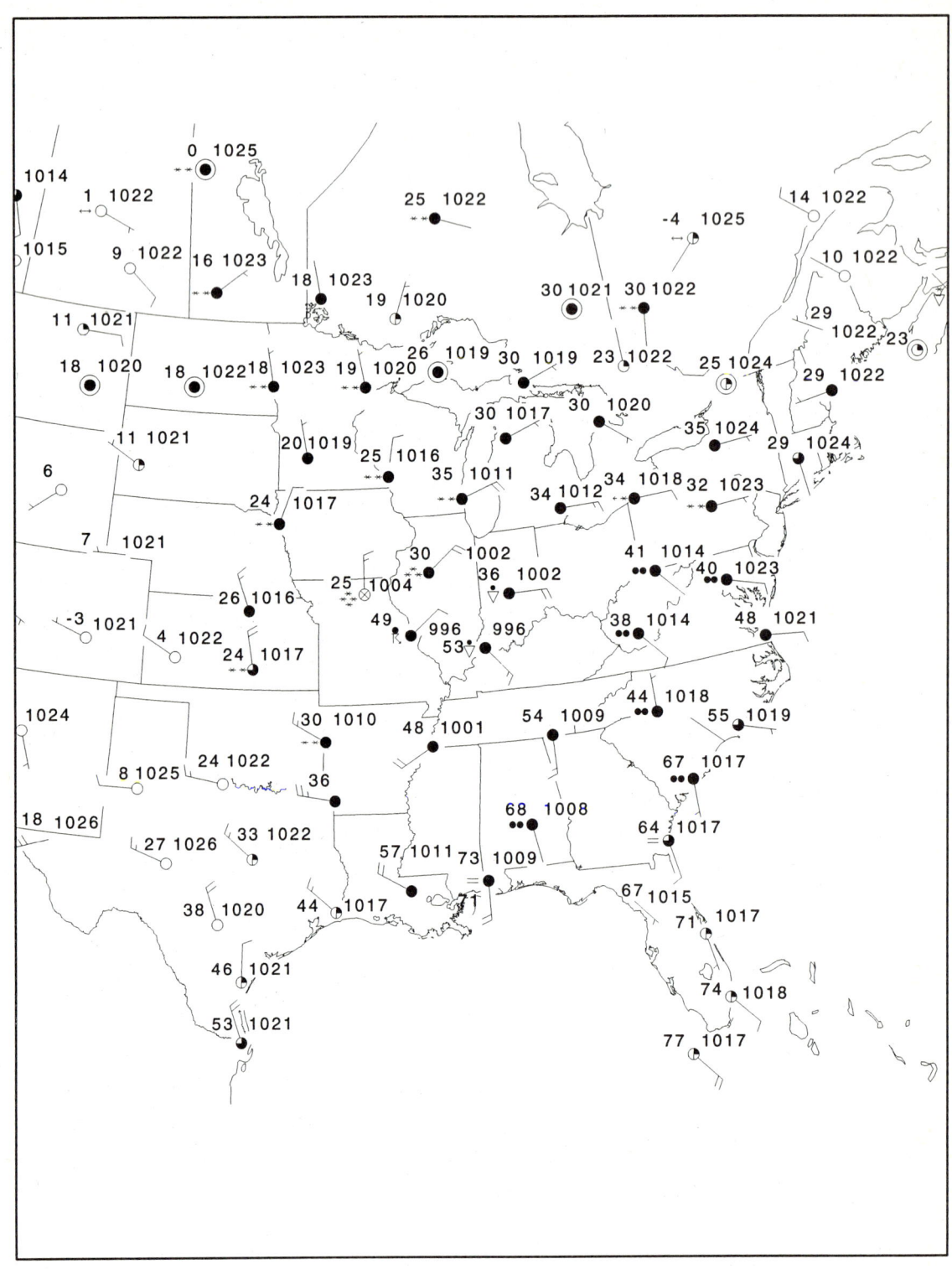

Figure 7.9b Plotted weather observations for 06 GMT 15 December 1987.

124

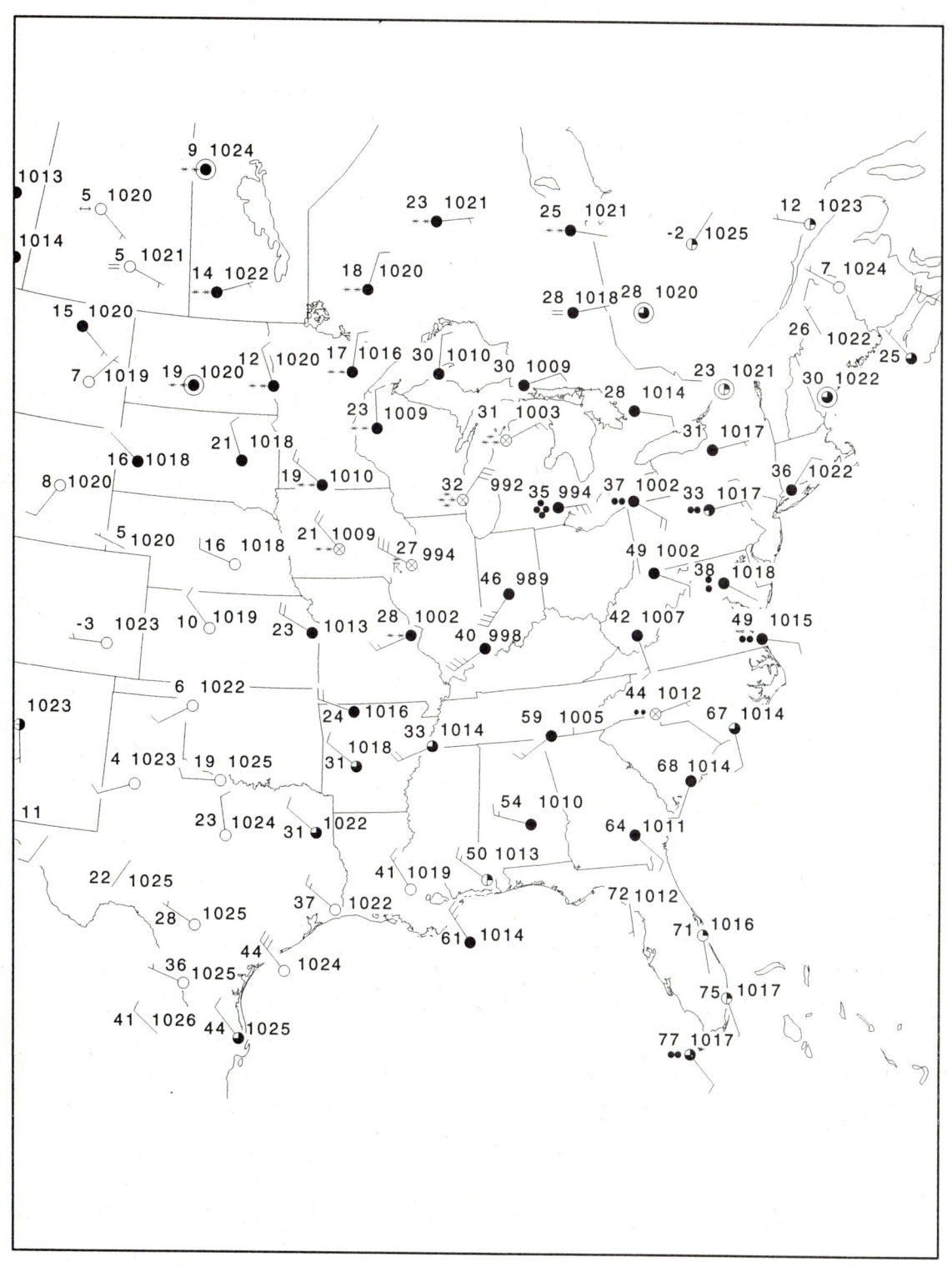

Figure 7.9c Plotted weather observations for 12 GMT 15 December 1987.

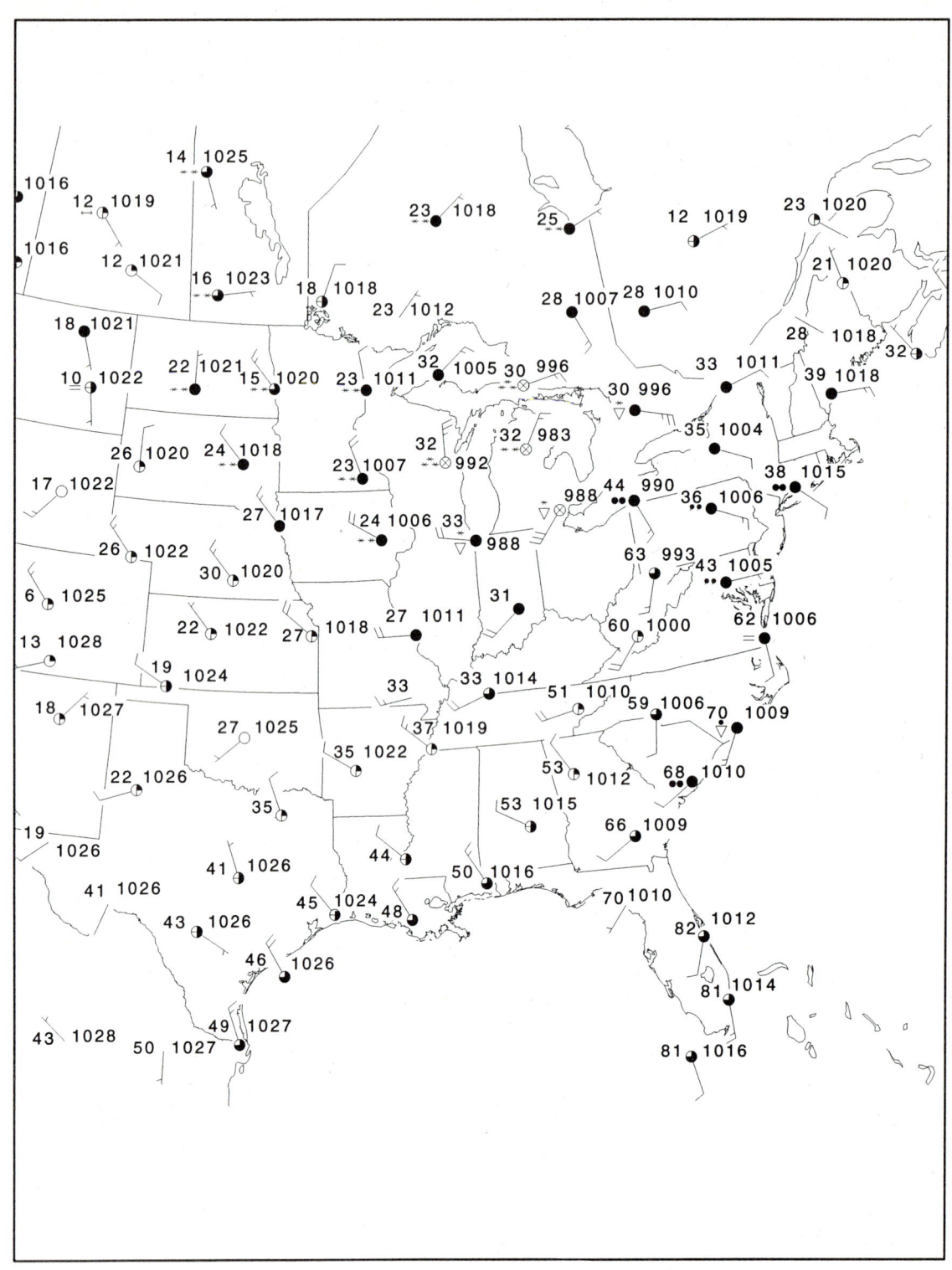

Figure 7.9d Plotted weather observations for 18 GMT 15 December 1987.

126

## QUESTIONS:

1. In Seattle the atmospheric pressure is 1010 mb. The temperature is 54°F. The dew point temperature is 50°F. The wind speed is 15 knots, from the southeast. The cloud cover is fifty percent and it is foggy. At the time of the last observation, the pressure was 1011.5 mb. Draw the station model for Seattle below.

2. Decode the following observation.

22  996
Moscow
16

3. Qualititively, what is the relationship between wind speed and spacing of isobars in your surface analyses?

4. What is the relationship between the isotherms and the position of the low pressure center?

5. What is the relationship between the isobars and the wind direction?

# Lab 28: Blizzards

## INTRODUCTION

"The Blizzard of '93" was a major winter storm that brought heavy snowfall along the Appalachians from Mississippi to Maine (Fig. 7.10). Electric power was lost in tens of thousands of homes and businesses due to the frozen precipitation and high winds. The death toll directly and indirectly related to the storm was estimated to be over 270 and the estimated cost of the storm is over $3 billion!

## ACTIVITY

OBJECTIVES: To goal of this activity is to analyze the relationship between the storm track and the hazardous weather it produced.

PROCEDURE:

1. On the accompanying map (Fig. 7.9) analyze the snowfall pattern by drawing contours of equal snow depth every ten inches (0, 10, 20, 30,..etc). Allow your contours to connect stations that are not so close together when it seems reasonable to do so in the overall pattern.
2. Use the list of locations for the storm center given in Table 7.1. to place an "L" with the day/time value for each position given in the Table.

**Table 7.1** A history of the central sea-level pressure and track of the storm.

| Day/Time | Latitude | Longitude | Pressure | Remarks |
|----------|----------|-----------|----------|---------|
| 12/ 7 pm | 28.2 N | 89.0 W | 989 mb | SSE of New Orleans, LA |
| 11 pm | 30.0 | 86.2 | 983 | SW of Pensacola, FL |
| 13/ 7 am | 32.0 | 83.0 | 973 | 30 mi. NW of Alma, GA |
| 1 pm | 35.5 | 78.5 | 966 | 30 mi. SE of Raleigh, NC |
| 7 pm | 38.7 | 75.8 | 960 | 30 mi. SW of Dover, DE |
| 11 pm | 40.9 | 74.3 | 962 | 20 mi. NW of LaGuardia, NY |
| 14/ 7 am | 45.0 | 68.1 | 965 | NE of Bangor, ME |

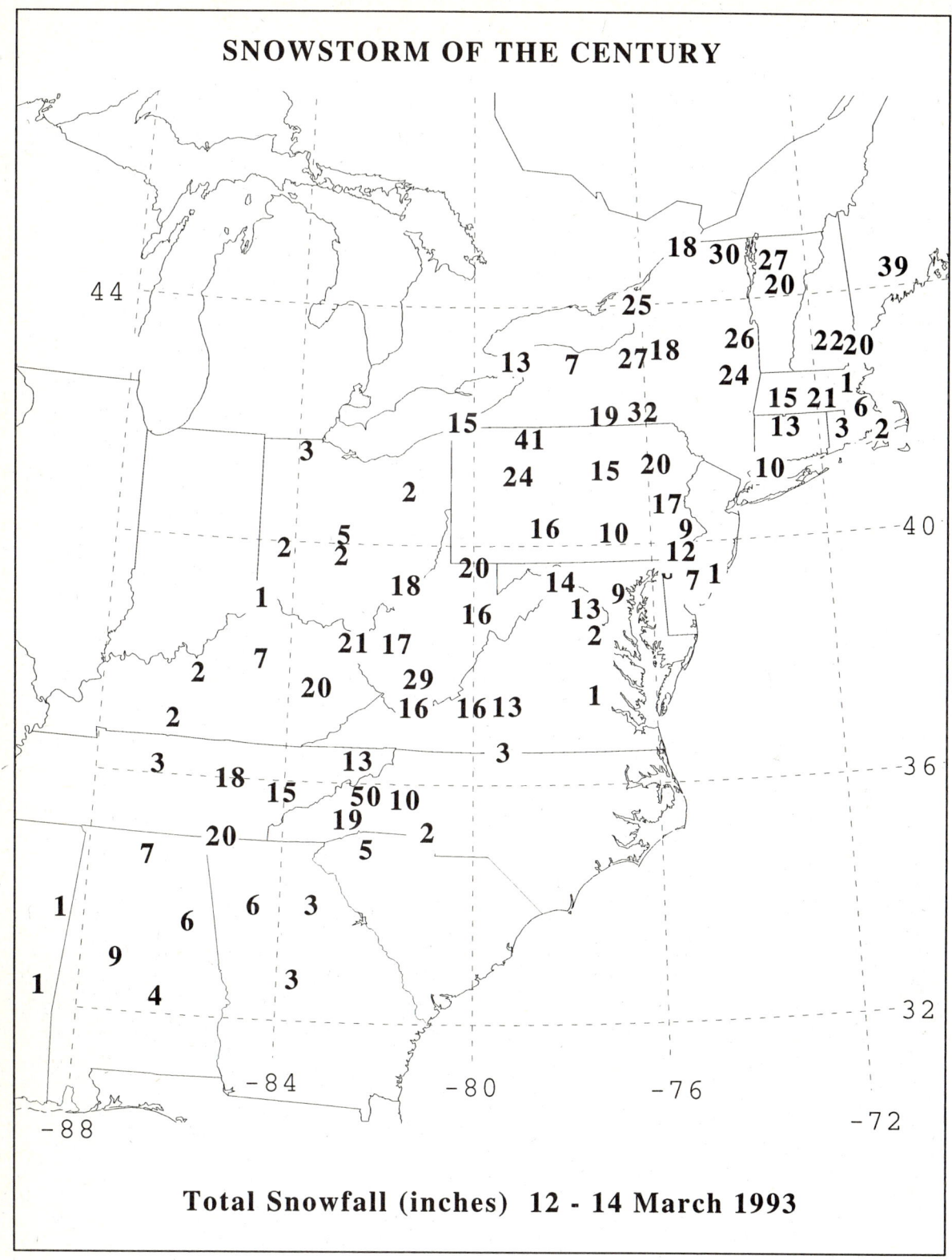

**SNOWSTORM OF THE CENTURY**

Total Snowfall (inches)  12 - 14 March 1993

Fig. 7.10

4. What is the relationship between the track of the cyclone center and the heaviest snowfall?   Explain this relationship in terms of the wind circulation and frontal structure of winter storms?  Why might you expect the coastal Carolinas to escape the snow in this case?

5. Note where the heaviest snowfalls occur in your analysis. What factors could account for the very high snow totals at some reporting stations?

6.  Have you ever experienced a snowstorm that was not well forecast in terms of expected snow accumulations? Note the large changes in snowfall totals between locations that are rather close to each other. Could this be a factor in the difficulty forecasters have in accurately predicting snow amounts?  What might account for such large variations over small distances?

22.   0040 GOES-7 IR 08 21 SEP 89264 200100 02153 08421 04.00

Vertical air speeds in the atmosphere usually are fairly small, being of the order of 10 cm/sec. In thunderstorms, however, the updrafts and downdrafts commonly exceed 10 m/sec (25 mph), and in severe storms they may be greater than 30 m/sec (75 mph).

Generally there is a balance between the upward-directed pressure-gradient force and the downward-directed force of gravity. This is called hydrostatic balance. When there is an imbalance between these two, vertical air motions arise. The difference between these two forces is called buoyancy or the buoyancy force. When a parcel of air is less dense than the surrounding air at the same altitude, the parcel is said to be buoyant and is subjected to an upward-directed force. At any height, the buoyancy depends mostly on the difference between the density of the parcel of air and the density of the surrounding environmental air. The density of the air depends mostly on temperature, but is decreased by the addition of water vapor and is increased by the addition of water drops or ice particles. For most purposes, the buoyancy force and upward acceleration of a parcel of air can be taken to be proportional to the difference between the parcel and the environmental temperatures.

The *stability* of the atmosphere is determined by the behavior of a parcel of air which is displaced from one altitude to another and then released. If it continues accelerating in the direction of the displacement, the atmosphere is *unstable*. If the parcel returns to the original altitude, the atmosphere is *stable*. If the released parcel remains at the new altitude, the atmosphere is neutral.

When a parcel of air ascends in the atmosphere, it moves to a region of lower pressure and expands. In the process, its temperature decreases because energy is used to do the work of expansion. In the absence of any external heat sources or sinks (such as latent heat of condensation), the rate of decrease of temperature of an ascending or descending volume of air is 10°C per kilometer, and this is known as the *dry adiabatic lapse rate*. Rising air cools, while sinking air warms at the same rate, 10°C/km.

The term *lapse rate*, which is widely used in meteorology, is the rate of decrease of temperature with increasing height. Thus, the dry adiabatic lapse rate is +10°C/km.

The stability of a cloudless atmosphere depends on the difference between the dry adiabatic lapse rate and the *environmental lapse rate*. The environmental lapse rate is the observed rate of change of the atmospheric temperature with increasing height. It normally is measured by means of a *radiosonde*, an instrumented package that is carried aloft by helium balloon. On average, through the lowest 10 km of the atmosphere, the environmental lapse rate is 6.5°C/km, but it varies widely from place to place and from time to time.

When the temperature increases with height, the lapse rate is negative, and a *temperature inversion* exists. In such a case, a rising parcel of air cooling at 10°C/km becomes increasingly cooler than the surrounding air. When this occurs, the atmosphere is considered to be stable because parcels return to their original levels.

Atmospheric stability is given according to the following differences between the dry adiabatic lapse rate (DALR) and the environmental lapse rate (ELR) (See Fig. 8.1):

ELR > DALR = unstable atmosphere

DALR > ELR = stable atmosphere

ELR = DARL = neutral atmosphere

When the atmosphere is unstable, there is enhanced vertical motion. One can visualize the upward and downward motion of air parcels or eddies. They transport such air properties as heat and pollutants from regions of high concentrations to regions of low concentrations. This process, known as eddy diffusion or turbulent diffusion, is important in mixing pollutants through the atmosphere. On a day when a temperature inversion is present, the atmosphere is stable, and there is little turbulent diffusion. As a result, pollutants released near the ground can become concentrated in a shallow layer. The behavior of plumes from smokestacks is governed to a large extent by atmospheric stability and the wind velocity.

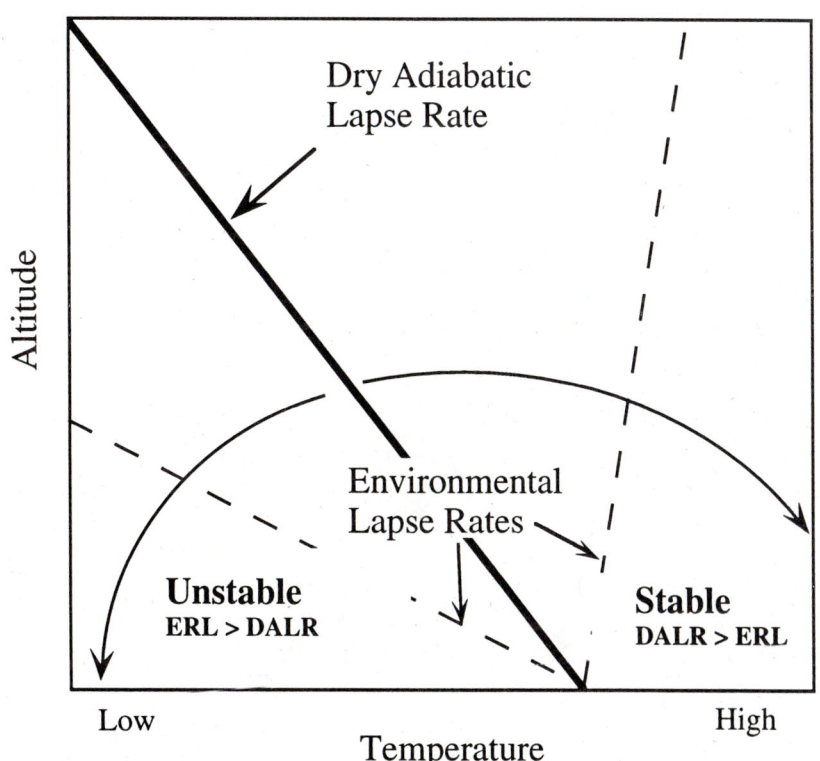

Figure 8.1 Stability diagram for dry air

The nature of vertical motion can be greatly influenced by the condensation and evaporation of water in the atmosphere. When air is moist, that is, when its relative humidity is high, relatively small upward displacements lead to saturation, condensation, and cloud droplet formation. This process causes the release of latent heat of condensation. This heat partially compensates for the dry adiabatic temperature reduction resulting from expansion. The rate of decrease of temperature of an ascending air parcel within which condensation is occurring varies, depending on the amount of water condensing, but it averages about 6°C/km. This temperature change is called the moist-adiabatic lapse rate. A descending parcel of air warms at the moist adiabatic rate if the air contains water drops which absorb heat as they evaporate while keeping the air saturated.

The latent heats released in the processes of condensation and freezing act to increase a cloud's temperature and increase its buoyancy. Thus the region of unstable lapse rates is greater under conditions of cloudy air than dry air (see *conditionally unstable* shaded area in Fig. 8.2). A process known as

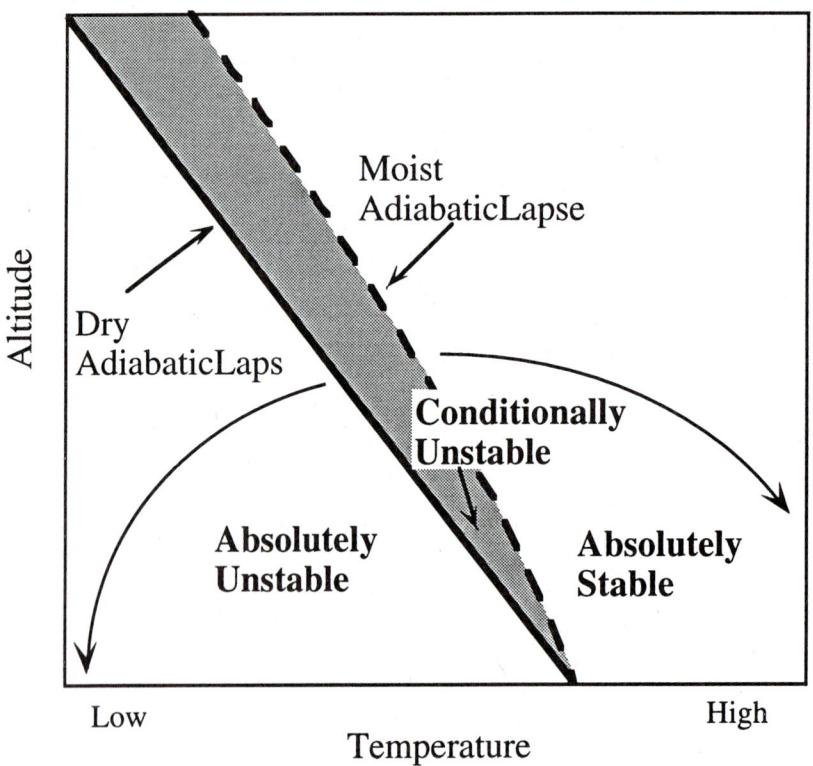

Figure 8.2 General stability diagram

entrainment has the opposite effect. Entrainment is the mixing of cloudy and non cloudy air across the cloud boundary. As the water and ice particles inside the cloud evaporate to saturate entrained dry air, the in-cloud temperatures are decreased. This reduces the cloud buoyancy. When clouds are relatively small and the surrounding air is very dry, entrainment can cause them to dissipate. Precipitation falling into dry air from elevated cloud bases can result in especially strong down drafts called *microbursts*.

### Severe Storms

The term severe storm commonly is used in reference to a weather system that is an immediate threat to life or property. Large, violent thunderstorms are the source of a great deal of dangerous weather. They can cause flash floods, yield lightning and hail, and on some occasions produce tornadoes.

Thunderstorms occur when the atmosphere is unstable and the air is sufficiently humid to be a fruitful source of latent heat. In some circumstances, thunderstorms can continue growing until they penetrate into the stratosphere, where stability inhibits further development. Thunderstorms are most frequent during the warm seasons of the year and, for the most part, are beneficial to most environments. Adequate thunderstorm rainfall during the growing season is crucial over the grain belts of the world. Unfortunately, when thunderstorms are unusually severe, the attendant winds and precipitation can cause widespread damage to most types of vegetation.

A thunderstorm is said to exist when a cloud produces lightning and thunder. There still is debate among the experts about the mechanisms by which a cumulonimbus cloud is electrically charged. Most authorities believe that the charging results from the interaction of ice particles and supercooled water drops within the updraft region of the cloud. It is known that the end result is a cloud containing mostly positive electric charge in its upper part and mostly negative charge in its lower part. A small center of positive charge is sometimes found in the rain under the cloud. When the accumulation of electric charges becomes sufficiently large to overcome the insulating properties of air, a huge electrical spark or arc occurs and is called lightning.

Thunderstorms often are classified as local storms or organized storms.

Storms in the first class (also called *air-mass* thunderstorms) tend to be small, less than 10 km in diameter, and short-lived. In less than an hour, they go through their entire life cycle of growth, maturity and dissipation. Such storms can yield a great deal of rain, but they seldom produce violent weather events other than lightning and moderately strong winds.

Organized thunderstorms generally are larger than local storms and may have durations of many hours. The most severe of the organized storms are called *supercell storms*. They tend to form in circumstances where dry, relatively cool air aloft moves rapidly over a layer of warm, humid air near the ground. When such a body of air is forced to rise, by an approaching cold front for example, the resulting instability of the air leads to a strong, persistent updraft that is tilted into the wind. Severe thunderstorms can produce large hail because some of the hailstones can be recycled through the updraft and spend much time colliding with and collecting supercooled water drops.

### Tornadoes

Tornadoes, probably the most feared of violent weather events, are usually associated with organized thunderstorms. Tornadoes appear as grey funnels, cylinders or ropes extending from a cloud base to the ground. Most often they are small, less than a few hundred meters in diameter, and are visible for only a few minutes. The most violent tornadoes, representing fewer than 10% of the storms (about 50 per year in the United States), do most of the harm. Tornado damage results primarily because of strong winds and secondarily because atmospheric pressure inside the funnel is much lower than that outside it. Tornadic winds that can exceed 100 m/s (224 mph) blow debris in all directions. When a funnel moves over a building, the pressure within the structure changes slowly relative to that outside it. The resulting difference in pressure leads to outward pressure forces that can contribute to the destruction of buildings by tornadoes.

Tornadoes occur in many countries, but not with the frequencies or intensities experienced in the United States. Tornadoes are most frequent in spring and early summer and are common over a broad area east of the Rocky mountains. The greatest occurEnce of destructive tornatoes is along a belt from central Texas across Oklahoma, Kansas, western Missouri, eastern Nebraska and into Iowa, and is known as "Tornado Alley."

The National Weather Service issues a *tornado watch* when a funnel is expected to form over a given region during the next few hours. When a tornado is sighted visually or by means of radar a *tornado warning* is issued. Communities in tornado-prone areas should adopt safety preparedness plans and conduct drills on how to react when a tornado watch or warning exists.

## *Hurricanes*

Hurricanes form over warm, tropical oceans and are the most destructive storms on Earth. In the western North Pacific, the storms that annually threaten the Philippines and Japan are called typhoons. In southeast Asia, the same type of storm is called a cyclone. Tropical cyclones with hurricane-speed winds develop over the Indian Ocean and sweep into countries with low lying coastal zones such as such as Bangladesh and India, with the resultS of extensive flooding and numerous fatalities. Despite their destructive potential, hurricanes also produce a great deal of beneficial rain. Tropical cyclones are not classified as hurricanes until their maximum wind speeds exceed 32.6 m/s (73 mph). In very strong hurricanes, the winds can exceed 100 m/s (224 mph).

A hurricane is characterized by a small central region known as the eye, within which the winds are light and there are few clouds. The eye is usually 20 to 40 km in diameter. Winds increase rapidly as one moves out of the eye and into a surrounding ring of thunderstorms referred to as the *eye wall*; maximum speeds are generally found in the eye wall at a distance of ~30 km from the center of the storm. A hurricane's overall diameter is typically between 500 and 1000 km.

There are five prerequisites for hurricane formation:

1) Warm ocean water - sea surface temperature > 26°C.
2) A pre-existing disturbance
3) Light winds aloft
4) Unstable troposphere (thunderstorm activity)
5) Large relative humidity in the middle troposphere

The energy source for hurricanes is the release of latent heat in the storm clouds. Given a source of moisture from a warm ocean surface, hurricanes typically last for a week to 10 days. The ones affecting the eastern United

States grow in intensity over the warm water of the Caribbean Sea and the Gulf of Mexico. The storms commonly are carried westward in the trade winds, but, as they approach the United States, they tend to curve toward the north and then northeast in part due to the influence of the prevailing westerlies at higher latitudes. When a hurricane moves over land or cold water, the supply of energy is reduced, and the wind speeds diminish. Over land, frictional forces also act to weaken the storm.

When a hurricane vortex moves over the ocean, the winds create large waves. Along coastlines, the winds cause an increase in the water level and flooding of low-lying coastal lands. Such a wind-induced, abnormal rise of the sea, called a *storm surge*, is historically responsible for most of the hurricane fatalities and damage. The effects of the storm surge are supplemented by heavy rains and strong winds. When hurricanes move over mountainous regions, the orographic lifting can lead to torrential rains and flooding, even at inland locations.

Weather satellites can effectively detect and track hurricanes over the entire Earth. When the storms get within a few hundred kilometers of land, radar and specially instrumented airplanes can also be used to observe a storm's intensity and the path it is following. When a hurricane is approaching a coastal location, early evacuation to higher ground of those people susceptible to the destructive force of the storm surge is essential. A *hurricane watch* is issued by the National Weather Service when there is a possibility of landfall within 36 hours. A *hurricane warning* is issued when landfall is likely within 12 to 24 hours. Large cities along the Gulf of Mexico are especially vulnerable to hurricanes and should be adequately prepared for them.

# Lab 29: Archimedes' Principle

## INTRODUCTION

Archimedes was a Sicilian scientist who was told by a king to find a way to determine whether or not a crown was made of pure gold. Some jewelers were not honest and would mix lead with gold when making crowns.

Archimedes was a hard working scientist. He often got so involved in solving problems and inventing things that he forgot to bathe. Then the king's soldiers would forcibly drag Archimedes off to the bath. According to the story, it was on one of these rare bath days that he learned how to distinguish between a fake crown and one of pure gold.

Archimedes noticed that the water level rose as he lowered himself into the bath. This meant that his body displaced a volume of water. Since an object's density is equal to its mass per unit volume, there must be a relationship between that object's density and the density of a fluid in which it is immersed. From this, we get Archimedes' Principle: an object immersed in a fluid will experience a buoyant force equal to the weight of fluid displaced by that object.

The direction of the buoyant force is up, opposing the weight of the object. Furthermore, if the object floats (i.e., its density is less than or equal to that of the fluid), the magnitude of its weight will be equal to the magnitude of the buoyant force. By measuring the mass of the king's crown in air and in water and comparing those measurements, the crown's density could be determined. Today, we have more efficient methods for determining the authenticity of crowns, but Archimedes' Principle is still used to study the behavior of objects in fluids.

For example, we know that fish are naturally equipped to deal with the buoyant force. Many fish have a sac called a swim bladder that is filled with gas. By releasing or taking in gas, fish can control their overall density and avoid being forced to the surface or to the ocean floor. This knowledge of how fish control density has aided development of exploratory vessels used in underwater research.

**ACTIVITY**

OBJECTIVE:  In the following activity, you will use Archimedes' Principle to predict what percentage of an iceberg is underwater.  You will then make measurements to test the accuracy of your predictions.

MATERIALS:
§ 2 half-gallon milk cartons
§ a tank or large bowl
§ fresh water (from the sink)
§ salt water (50 grams of table salt to 1 liter of water)
§ measuring tape

PROCEDURE:
The day before doing this procedure, pour enough water to fill the bottom third of two half-gallon milk cartons.  Freeze.

The density of an object is its mass per unit volume.  The density of ice is $0.92$ g/cm$^3$ .  The density of fresh water is $1.00$ g/cm$^3$.  The density of salt water (35 ‰) is $1.03$ g/cm$^3$.

Since ice is less dense than fresh water or salt water, it will float.
The density equation is  $\rho = m/V$ or $m = \rho V$ ; since $W = mg$, then $W = \rho V g$.

   $\rho$ = density          $m$ = mass          $V$ = volume
   $W$ = weight          $g$ = gravity

1. Write an equation relating the buoyant force to the water's density, volume and acceleration due to gravity.



2. Write an equation relating the weight of an iceberg to its volume, density and acceleration due to gravity.

3. Using the two equations above, find the percentage of an iceberg that is below the surface of fresh water.


Repeat Step 3 for salt water.
Remove your icebergs from the milk cartons, and measure the volume.

4. Volume of iceberg in fresh water (Iceberg 1) = _____

5. Volume of iceberg in salt water (Iceberg 2) = _____

Very gently, place Iceberg 1 into a tank or large bowl of fresh water. As quickly and as accurately as possible, scratch out the line where Iceberg 1 breaks the surface of the water.

6. Measure the volume of Iceberg 1 that is above the surface of the water. (This part doesn't melt as much as the submerged part. What is the reason for this?) Subtract this number from the total volume of Iceberg 1 found in Step 4. Compute the percentage of Iceberg 1 that is below the surface.


Repeat the above procedure, placing Iceberg 2 in salt water.
Repeat the calculations in Step 6 for Iceberg 2.

7. Percentage below surface (fresh water) _____

8. Percentage below surface (salt water) _____

9. Compare your measured values with those you predicted earlier. Use percent difference:

   % difference = _____ measured value _____ x 100 measured value - predicted value

10. Finish the drawing of the iceberg shown below.  Sketch in the bottom of
    the iceberg, giving an indication of how much ice is above and below the
    surface of the water.

Fig. 8.3

# *Lab 30: Convection*

## INTRODUCTION

Nearly all of the heat energy and moisture contained in the troposphere was input at the Earth's suface through a combination of conduction and turbulent mixing. This is because incoming sun light largely passes through the atmosphere unimpeded and is absorbed at the Earth's surface. Since the atmosphere is heated and moistened from below, convective currents redistribute this heat and moisture up through the troposphere. Stronger updrafts are visible as towering cumulus clouds and thunderstorms. Thunderstorm activity is especially prevalent in satellite imagery over land in the tropics (e.g., Amazon River Basin, Congo), where the solar radiation and surface heating are a maximum.

## ACTIVITY

OBJECTIVE: The purpose of this activity is to investigate and observe how material moves within a convection cell. This information will be used to gain a better understanding of how clouds and air circulations evolve and redistribute energy in the Earth's atmosphere.

MATERIALS:
§ cafeteria tray
§ white paper
§ three styrofoam cups
§ glass pan
§ water pitcher
§ food coloring
§ pipet or eyedropper

PROCEDURE:
1. Line the cafeteria tray with white paper to make it easier to observe the flowing food coloring.
2. Place three styrofoam cups upside down on the paper. The fourth cup eventually will be placed right side up between the other three.
3. Fill the pan one-half to two-thirds full of water from one of the pitchers.

4. Place the pan on the inverted styrofoam cups as shown below.
5. The water in the pan should sit for a minute or so before any food coloring is added to be sure there are no initial currents in the water.

Fig. 8.4

*Trial 1*

Observe the movement of the food coloring when there is no heat source present. This will provide a basis for comparing the various trials which do employ a heat source.

6. After the water in the pan has had an opportunity to sit undisturbed for a minute, place a very small drop of food coloring in the center of the pan. The food coloring should be placed on the bottom of the pan. Move the pipet straight up and down in the pan of water and take care not to create currents in the water. Slowly release the drop of food coloring when the tip of the pipet touches the bottom of the pan, so as not to unduly disturb the circulation of the water.

7. On the data sheet, draw what happens to the food coloring as you look both from the top and from the side view. You may need to hold a piece of white paper behind the pan in order to see the food coloring better when viewing from the side. Write a brief description of what you see in the space provided on the data sheet.

8. Once you are satisfied that you have a clear understanding of the movement of the food coloring in this trial, gently swirl the water to disperse the food coloring. The water in the pan need not be changed after the first trial, unless the water becomes too dark to be able to observe movement of additional food coloring placed in the pan.

144

*Trial 2*

Observe the movement of food coloring when a heat source is placed directly underneath the center of the pan. As in Trial 1, the food coloring will be placed on the bottom, in the center of the pan.

9. Get a styrofoam cup of hot water. Take care to avoid spilling the hot water. Carefully slide the cup underneath the center of the pan of water, and allow the water in the pan to become still.

10. Follow the steps as in Trial 1. After completing Trial 2 replace the water in the glass pan with cool clear water and refill the cup with hot water. Describe the results in the Data Table.

*Trial 3*

Observe the movement of food coloring with the heat source placed under the center of the pan and with the food coloring placed on the bottom and to one side of the center of the pan.

11. Follow each of the steps outlined in Trial 2, but for this trial, place a small drop of food coloring at the bottom of the pan about halfway between the center of the pan and the side. Remember to slowly release the food coloring when the tip of the pipet touches the bottom of the pan so as not to unduly disturb the circulation of the water. Describe the results in the Data Table.

*Trial 4*

Observe the movement of food coloring with the heat source placed under the center of the pan and with the food coloring placed to one side of the center of the pan on the top of the water.

12. Again, follow each of the steps outlined in Trial 2, but for this trial, place a small drop of food coloring about halfway between the center of the pan and the side. Place the drop of food coloring directly on the top of the water rather than on the bottom of the pan. Describe the results in the Data Table.

## Data Table

| | TOP | SIDE VIEW |
|---|---|---|
| **TRIAL 1**<br>heat = none<br>color = bottom, center | | |
| **TRIAL 2**<br>heat = center<br>color = top, center | | |

## Data Table

| | TOP | SIDE VIEW |
|---|---|---|
| *TRIAL 3*<br>heat = none<br>color = bottom, side | | |
| *TRIAL 4*<br>heat = center<br>color = top, side | | |

QUESTIONS:

1. Summarize the results in the data tables from the four trials by describing a convection cell in general terms. Draw a schematic model of a convection cell, showing the heat source and fluid motions.

2. If the water in the glass pan represents the atmosphere, what does the hot water in the styrofoam cup represent?

3. The currents of water cause the food coloring to move at a rate of 2 to 3 cm or more per minute. By observing the growth of cumulus clouds on a fair day, how quickly would you say the air circulates through the clouds? How do these two figures compare?

4. Convection is a process which not only occurs in the atmosphere, but is also the driving force for other phenomena. Relate some other areas in which convection is important.

# Lab 31:  Tracking Hurricane Hugo

## INTRODUCTION

Because of their destructive nature, it is very important to accurately track and forecast the movement of hurricanes.  The location of the center of a hurricane is obtained through a combination of observations from satellite, aircraft, ships, and islands.  The motion thus obtained can then be extrapolated into the future and compared with computer simulations of the storm track in order to make predictions of the future path and speed of the storm.  Watches and warnings are provided to the public by the National Weather Service based on the expected storm track to give people in affected areas time to prepare and evacuate as needed.

## ACTIVITY

OBJECTIVE: The objective of this activity is to follow the track of hurricane Hugo for several days, with the goal of anticipating the storm's landfall. A second objective is to learn the meaning of  hurricane watches and warnings.

> A *hurricane watch*  is issued by the National Weather Service when there is a possibility of landfall within 36 hours.

> A *hurricane warning* is issued when landfall is likely within 12 to 24 hours.

MATERIALS:
§ pencil
§ map

PROCEDURE:
1. Study the information given in Table 1.  It contains information collected from hurricane Hugo.  In it there are three types of data: a) Date/Time, b)Position- the location of the eye of the hurricane, c) Wind Speed- this is the speed of the maximum winds in the hurricane (not the speed of the hurricane's motion).
2. Plot the data in Table 8.1 on the map provided in Fig. 8.5.

Table 8.1 Location of hurricane Hugo and maximum sustained winds.

| DATE/TIME | POSITION | | WIND |
| | Lat. (°N) | Lon.(°W) | SPEED (kts) |
|---|---|---|---|
| 14/0000 | 13° | 45° | 70 |
| 0600 | 13 | 46 | 80 |
| 1200 | 13 | 48 | 85 |
| 1800 | 14 | 49 | 90 |
| 15/0000 | 14 | 50 | 100 |
| 0600 | 14 | 52 | 110 |
| 1200 | 14 | 53 | 125 |
| 1800 | 15 | 55 | 140 |
| 16/0000 | 15 | 56 | 135 |
| 0600 | 15 | 57 | 130 |
| 1200 | 15 | 58 | 120 |
| 1800 | 16 | 59 | 120 |
| 17/0000 | 16 | 60 | 120 |
| 0600 | 16 | 61 | 120 |
| 1200 | 17 | 62 | 125 |
| 1800 | 17 | 64 | 125 |
| 18/0000 | 17 | 64 | 130 |
| 0600 | 18 | 65 | 120 |
| 1200 | 18 | 66 | 110 |
| 1800 | 19 | 66 | 105 |
| 19/0000 | 20 | 67 | 100 |
| 0600 | 21 | 67 | 90 |
| 1200 | 22 | 68 | 90 |
| 1800 | 23 | 69 | 90 |
| 20/0000 | 24 | 69 | 90 |
| 0600 | 24 | 70 | 90 |
| 1200 | 25 | 71 | 95 |
| 1800 | 26 | 72 | 95 |
| 21/0000 | 27 | 73 | 100 |
| 0600 | 28 | 75 | 100 |
| 1200 | 29 | 76 | 110 |
| 1800 | 30 | 78 | 120 |
| 22/0000 | 32 | 79 | 120 |
| 0600 | 34 | 80 | 85 |
| 1200 | 36 | 82 | 55 |
| 1800 | 38 | 85 | 40 |
| 23/0000 | 42 | 80 | 35 |
| 0600 | 46 | 74 | 40 |
| 1200 | 49 | 69 | 40 |
| 1800 | 51 | 65 | 40 |

QUESTIONS:

1. Based upon the National Weather Service's definition of a hurricane warning, at what time should one be issued for San Juan, Puerto Rico, and for Charleston, SC? What was its speed of motion at these two times?

2. In which locations did Hugo appear to do the most damage before it hit South Carolina?

3. What was the general direction of the storm motion of Hugo prior to hitting South Carolina? Since the strongest winds are located on the right-front side of hurricanes. Indicate on Fig. 8.5 with a red "X" where you expect the greatest damage be?

4. What happened to the storm speed and direction of Hugo after crossing the South Carolina coast? Why?

5. Judging by the wind speed, indicate on Fig. 8.5 with blue "X"s where Hugo was first upgraded to a hurricane and where it was downgraded from a hurricane to a tropical cyclone?

6. The strongest hurricanes produce slower wind speeds than the strongest tornados, but a hurricane inflicts more total damage. How might this be explained?

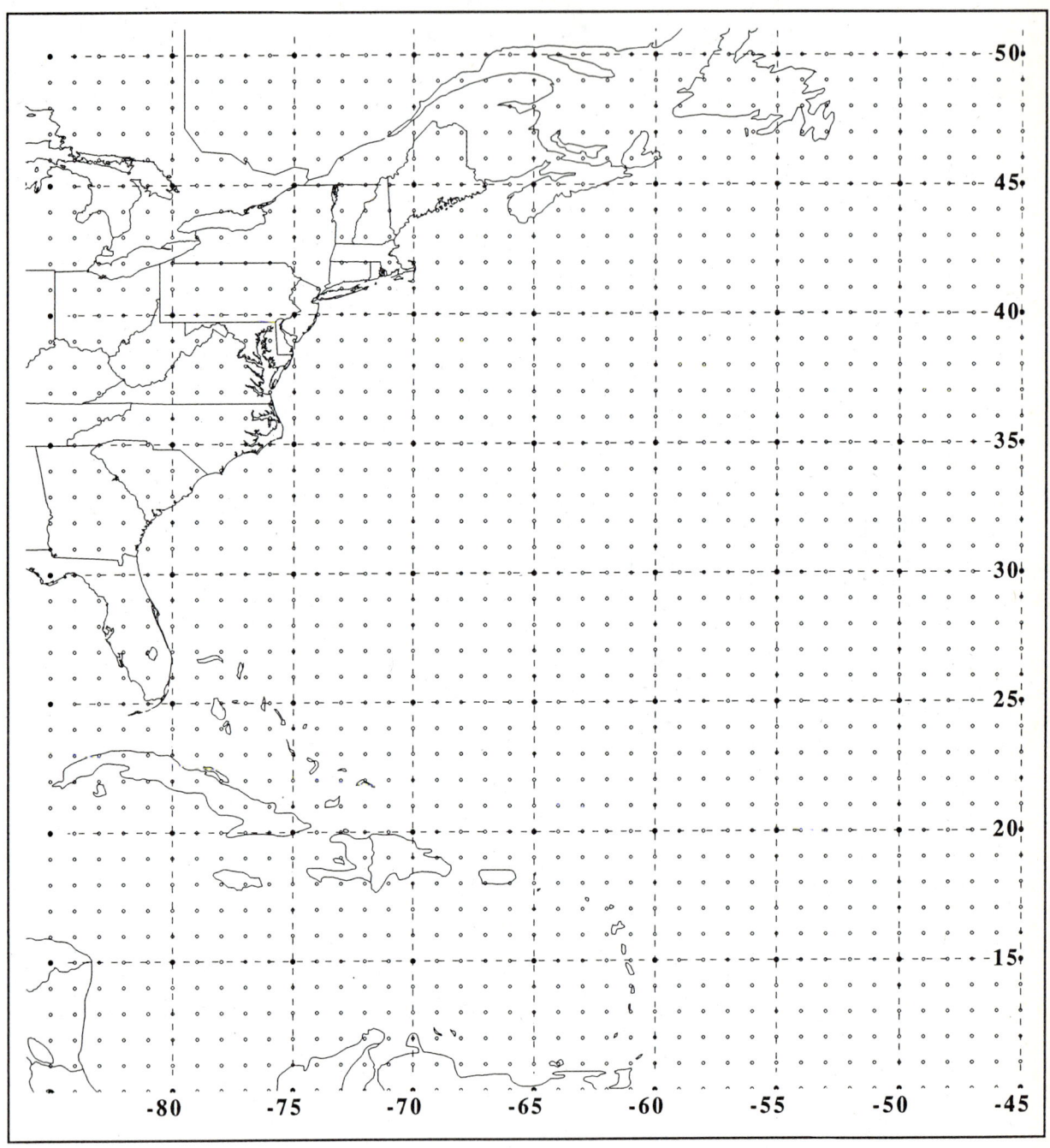

Figure 8.5

# Lab 32: Weather Radars as Forecasting Tools

## INTRODUCTION

Radar (short for radio detection and ranging) was developed during World War II after it was discovered that electromagnetic waves emitted by an antenna on the ground could be reflected off an aircraft, thereby allowing the aircraft to be detected remotely. As the power and sensitivity of military radars increased, precipitation began to appear on radar screens and the weather radar was born. Since then weather radars have become one of the most important tools in weather forecasters' arsenal. The strength of the signal reflected by precipitation particles gives information on the rate of rainfall (Table 8.2) and on the structure of storm systems (Fig. 8.6)

Modern radars, like the ones recently deployed by the National Weather Service, not only detect precipitation rates but also measure the component of velocity of the precipitation particles in the direction parallel to the radar beam. The ability to measure velocity comes from analyzing the phase or Doppler shift of the returned signal, and is the same principle used by police radar to gauge the speed of cars on the highway. This added capability is very important for measuring rotation in potentially tornadic thunderstorms so that more timely warnings can be issued in cases of severe weather.

## ACTIVITY

OBJECTIVE: The objective of this activity is to become familiar with the use of radar reflectivity data in storm analysis and forecasting.

PROCEDURE:
Use the information in Table 8.2 below to interpret the radar data for hurricane Hugo given in Figs. 8. 6 a-d, then answer the questions that follow.

Table 8.2 Reflectivity Intensity

| D/VIP Level | Echo Intensity | Estimated Precipitation | Rainfall Rate (inches/hour) Convective |
|---|---|---|---|
| 1 | Weak | Light | < 0.2" |
| 2 | Moderate | Moderate | 0.2 - 1" |
| 3 | Strong | Heavy | 1 - 2" |

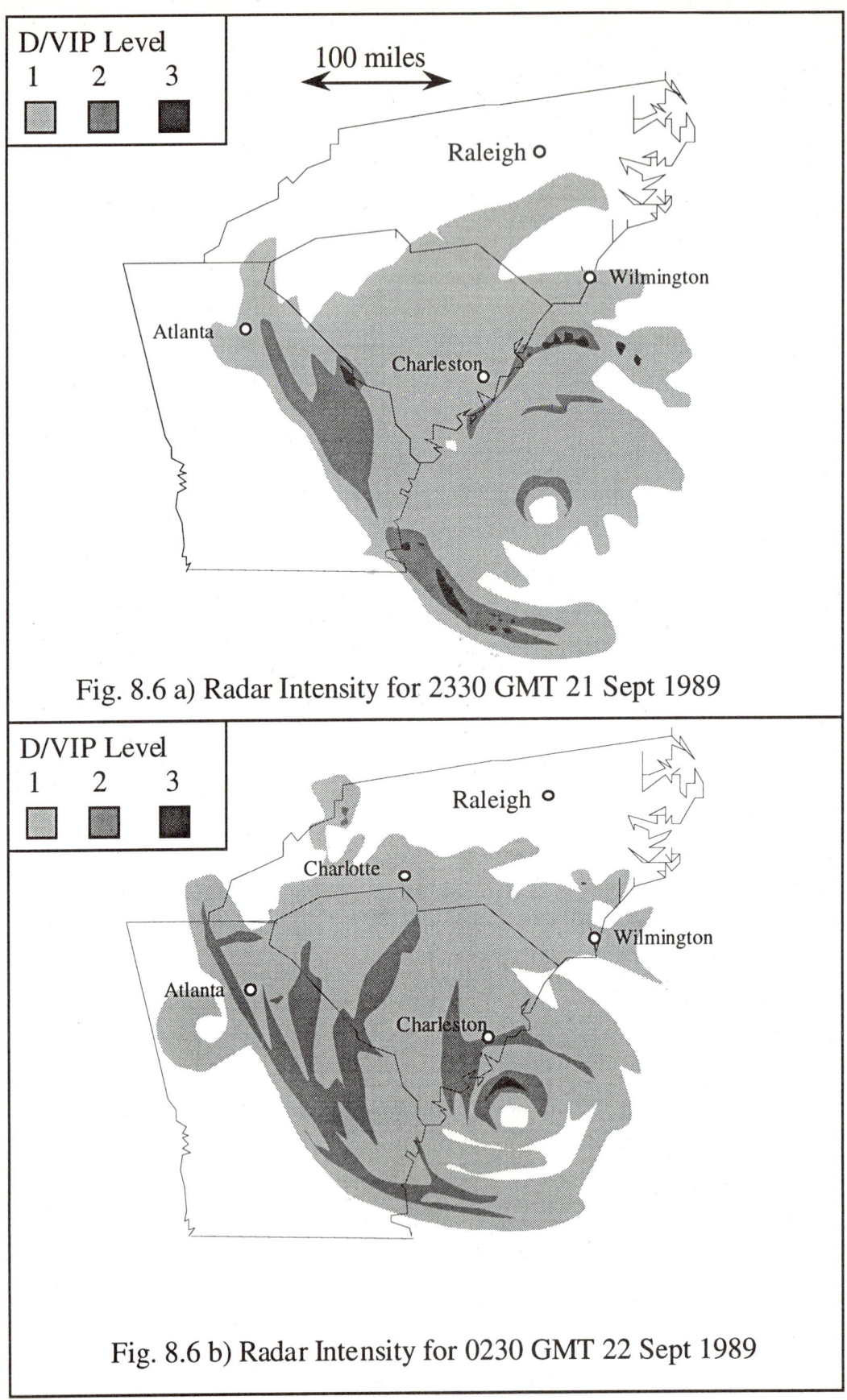

Fig. 8.6 a) Radar Intensity for 2330 GMT 21 Sept 1989

Fig. 8.6 b) Radar Intensity for 0230 GMT 22 Sept 1989

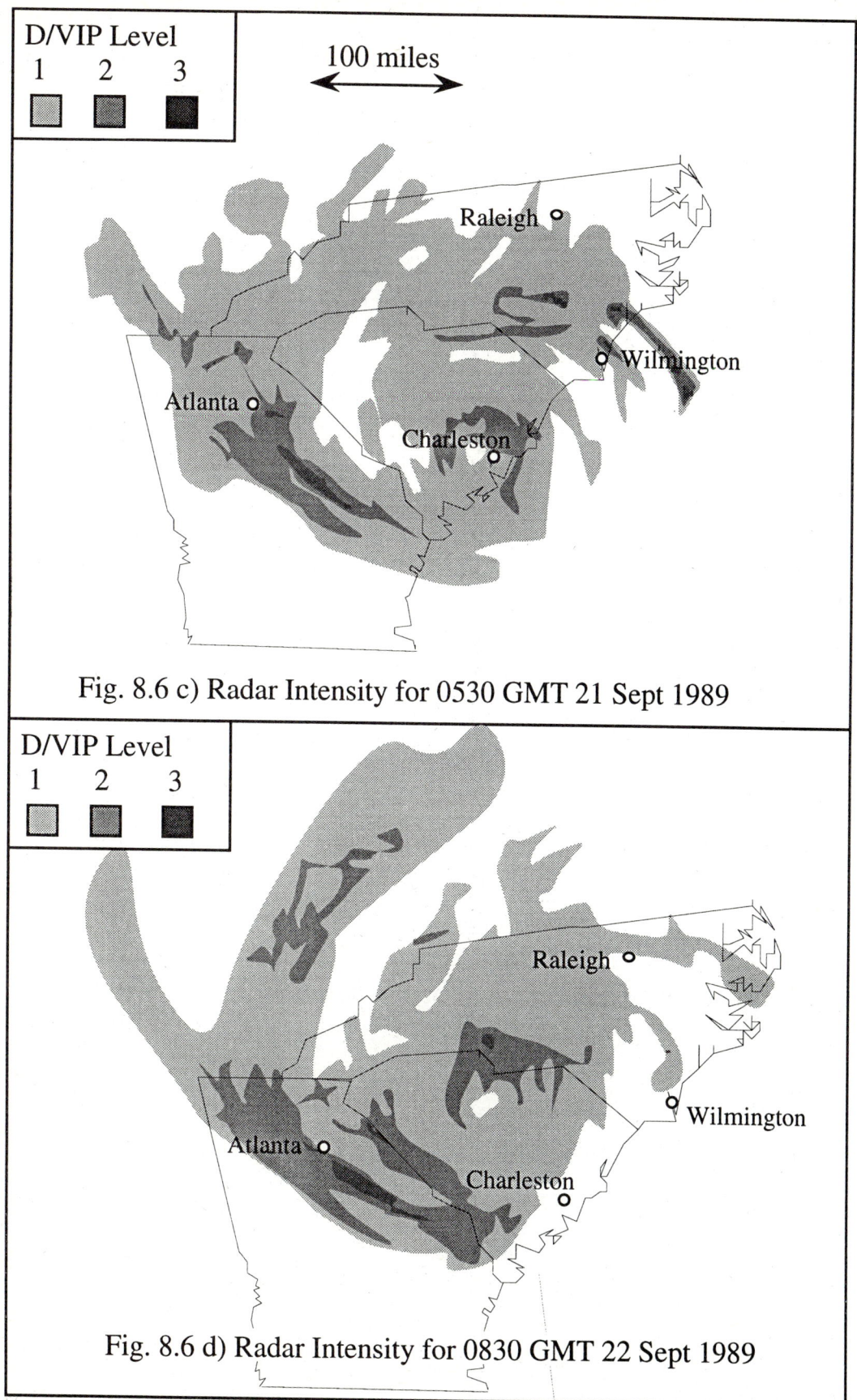

Fig. 8.6 c) Radar Intensity for 0530 GMT 21 Sept 1989

Fig. 8.6 d) Radar Intensity for 0830 GMT 22 Sept 1989

Look at Fig. 8.6 which shows the rainfall rate for hurricane Hugo at four times to answer the questions below. The times are given in Greenwich Mean Time (GMT) which is the time standard used by meteorologists to deal with the problem of time zones when collecting and analyzing weather data across large distances.

1. Locate the signature of Hugo's eye and eyewall. Based on these and the distance reference provided in Fig. 8.6, calculate the propogation speed of Hugo for each time interval. What affect does landfall have on the speed?

2. How does Hugo's precipitation structure change after landfall?

3. Hurricane winds and storm surge are greatest in the eyewall on the right side of the storm when looking the direction of storm motion. Based on the radar intensity data where do you expect the greatest damage from these hazards occurred with Hugo?

4. On Fig. 8.7 below, sketch contours of storm total rainfall in one inch increments based on the individual panels in Fig. 8.6 and the fact that the maximum storm total precipitation was a little more than 5 inches? To estimate which areas are most prone to flooding you need to look not only at the rainfall intensity, but also at the speed with which precipitation features move across an area. Rainbands that are relatively slow moving or stationary contribute to local flooding. Compare you estimate to the pattern given in Fig. 8.8.

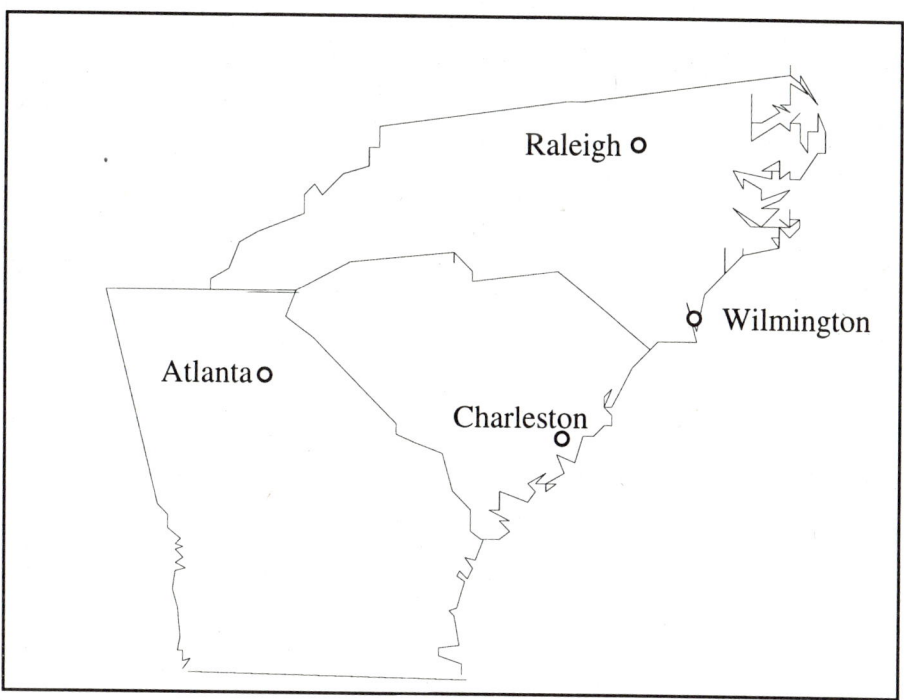

Fig. 8.7

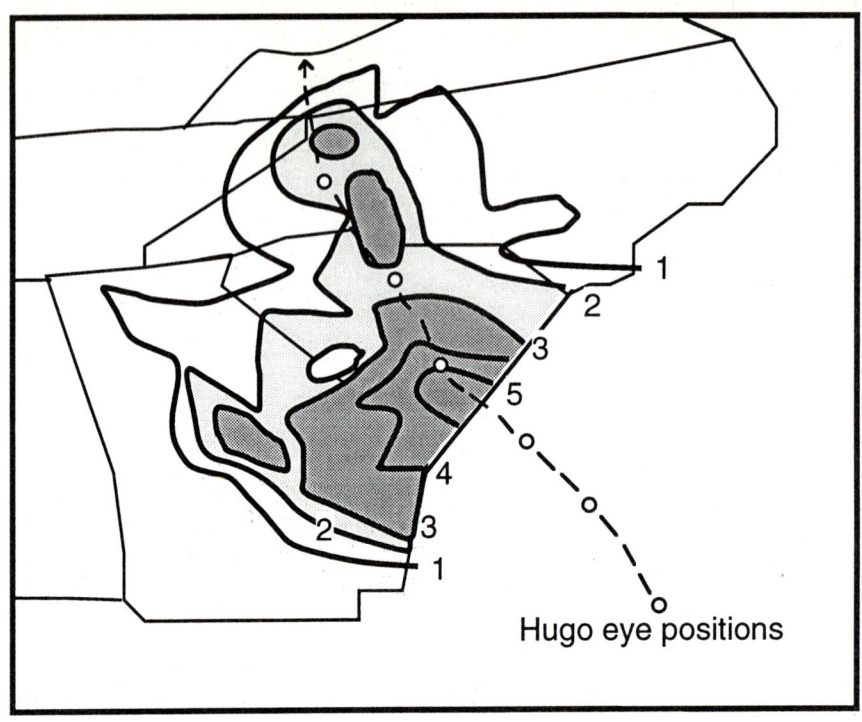

Fig. 8.8 Storm-total precipitation (inches) for the 19-h period from 21 September 1989 at 1900 GMT to 22 September at 1300 GMT. Light shading represents rainfall > 2 inches and dark shading represents rainfall > 3 inches. The eye positions are in 3-h increments starting at 2100 GMT.

# *Lab 33: Wind Generated Waves*

## INTRODUCTION

Wind generated waves on lakes and oceans represent a serious hazard to maritime interests. When they impact the coast, waves can produce serious erosion and flooding problems even far from the disturbance that initially generated the waves. The highest wind-produced wave ever recorded was 34 meters (112 feet) high. In 1933, the U.S. Navy tanker, *U.S.S. Ramapo*, was traveling from the Philippines to San Diego. During transit, the tanker encountered a storm system that produced strong winds (up to 70 mph) for more than a week across the Pacific Ocean. At the height of the storm, observers on the ship saw a mountainous wave that crested even with a platform on the crow's nest mast. From the height of the tanker's bridge, mast and stern, it was calculated that the wave was 34 meters high--as high as a nine-story building. Such immense waves are sometimes refered to as "rogue" waves.

The wind-generated ocean waves are sometimes mistakenly called tidal waves or "tsunami," Japanese for harbor wave, but the cause of a tsunami is seismic activity, such as an underwater Earthquake not wind. One of the largest tsunami ever recorded was 93 meters (278 feet) high and was sighted off Ishigaki Island, Ryukyu Chain on April 24, 1971. It moved a block of coral, weighing 850 tons, 1.3 miles.

Wave concepts have broad application in the geosciences. For example, measuring wind speed using a Doppler radar involves applying wave concepts, as does understanding the dynamics of gravitiy waves in the atmosphere.

## ACTIVITY

OBJECTIVE: In this activity, the frequency, wavelength, and speed of water waves will be analyzed. The results should demonstrate that a wave transfers energy, not material. The concepts of reflection, refraction, diffraction, and interference will also be explored.

## MATERIALS:

§ ripple tank (if demonstrating only)
§ large, flat cake pans or cookie sheets
§ portable fan
§ watch or clock with a second hand
§ rulers or any blunt, straight object
§ large chunks of clay

## PROCEDURE:

1. Fill your ripple tank with water. A depth of ~1/2 inch works well. Allow the water to settle, and use a portable fan to generate waves (Fig. 8.8).

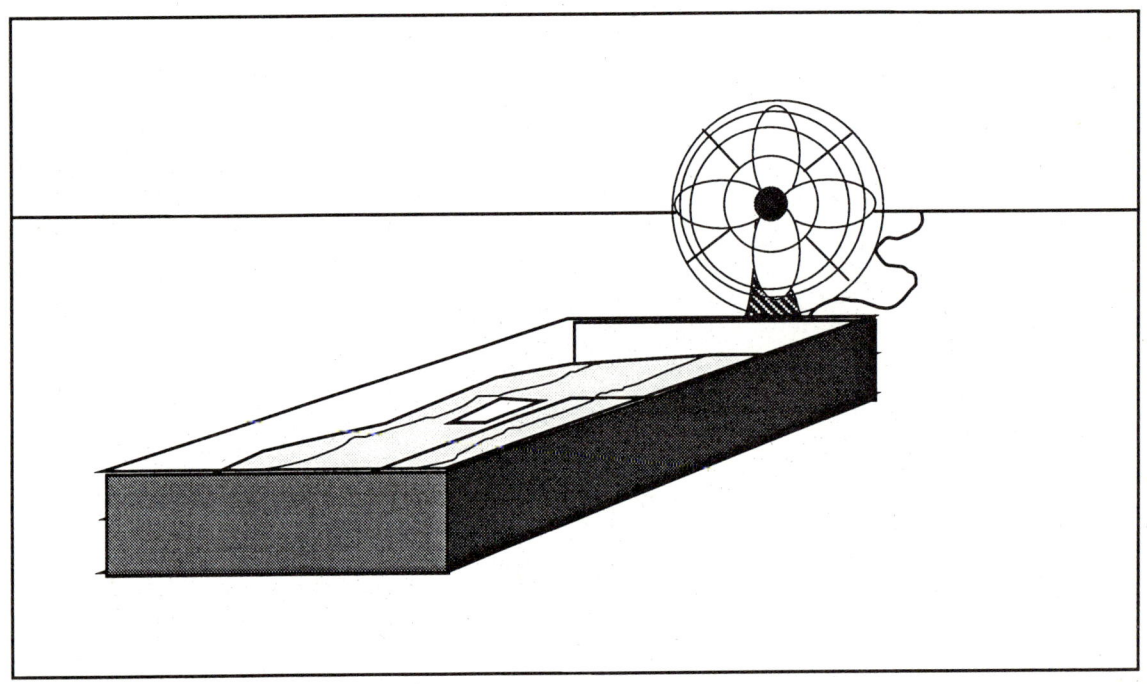

Fig. 8.8

2. Notice how the waves reflect off of the sides of the tank. Can you see how the incident or incoming waves interfere with the reflected waves? It occurs quickly so you will need to look carefully. The waves appear to move right through each other.

3. Now place a tiny piece of paper in the water. Let the water settle. Then send a wave from one end of the tank toward the paper. The paper should just vibrate back and forth as the wave passes, but otherwise stay put.

4. This demonstrates that it is energy that is moving across the tank and not individual water molecules. A wave is just a pulse of energy traveling from one place to another.

5. Finally, place a chunk of clay in the water to act as a model island. By making waves that strike the "island," you can simulate the way ocean waves diffract as they pass islands. If you fill (or empty) the tank so that the clay islands become slightly submerged, you can simulate the way ocean waves refract as they pass over reefs and sand bars.

QUESTIONS:

Look at Fig. 8.9 which is a time plot showing the number of waves passing a fixed point B in 2.5 secs. The distance between the wave crests (wavelength) is 2 cm. Using this information, answer the following questions.

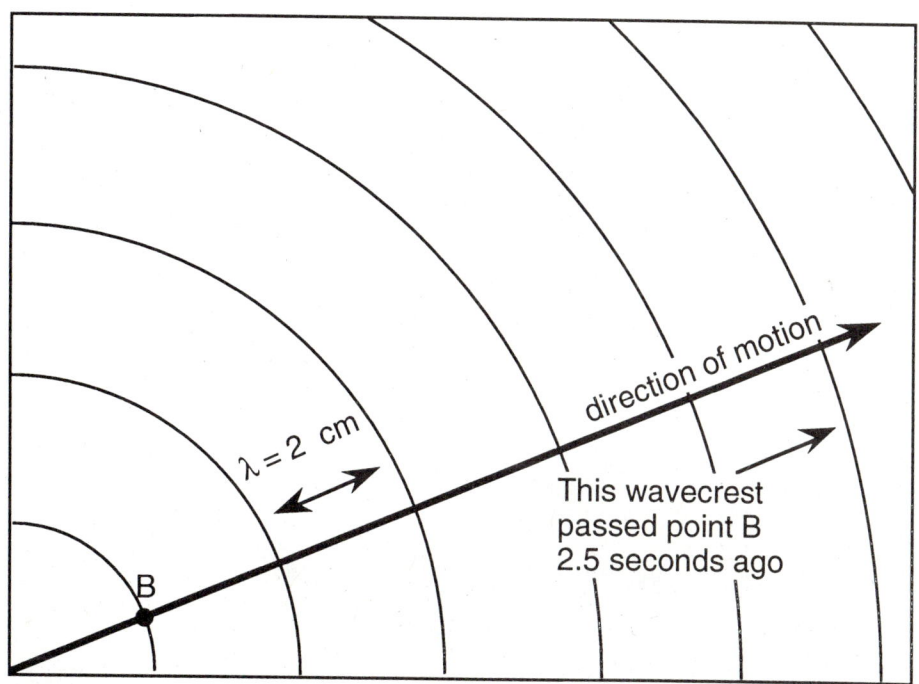

Fig. 8.9 Waves passing a fixed point B.

1. How many waves will pass point B (or any point) each second? This is the frequency of the waves in units of waves per second or Hertz.

2. How far will the first wave to pass point B have traveled after one second has gone by? (Hint:  What does the wavelength represent?)

3. Using the fact that speed = distance x time, calculate the speed of the water wave.

4. Notice that the same result can be obtained using the equation:
   speed = frequency x wavelength.     Why does this work?

5. Draw a sketch in the space below to help demonstrate how the tanker crew used the physical dimensions of their ship to calculate the height of the largest wave ever sighted.  The *U.S.S. Ramapo* had a length of 164 meters.  The crow's nest is located midship with a height of 17.5 meters. The wave had a period (P) of 14.8 seconds.

6. What is the length ($\lambda$) of the wave? [Use this formula: $\lambda(m) = 1.56$ $P^2(sec)$.]

7. Using this $\lambda$ and length of the ship, how could you determine the height of the wave? The observer is on the bridge.

Air pollution is composed of airborne substances (either solid, liquid, or gaseous) that threaten the health of the Earth's flora and fauna (including animals and people). The source of air pollution can be natural (forest fires, volcanic eruptions, wind born dust, etc.) or through human activities (burning of fossil fuels, release of refrigerants, etc.). By far the majority of air pollution is released into the atmosphere at the Earth's surface. Therefore, in the absence of winds and upward mixing, pollution can rapidly accumulate, especially in urban areas, leading to pollution episodes. The term "smog," a combination of the words smoke and fog, was coined to describe such events. Cold, stable air drains into basins and valleys from adjacent higher terrain, especially during the longer nights of winter, causing suppressed air circulation in these regions. In December 1952, a particularly bad smog episode in London claimed the lives of over 4000 people. Since that time legislation, such as the Clean Air Act in the U.S., has reduced air pollution levels in urban areas, decreasing the threat of such disasters in developed nations.

A major pollutant of city air is carbon monoxide, a colorless, odorless gas emitted by internal combustion engines (primarily cars and trucks). Carbon monoxide is actively taken up by our respiratory systems displacing oxygen intake and leading to suffocation when the concentrations are too high. There has been about a 40% decrease in carbon monoxide in the U.S. since 1970, due to stricter air quality standards and the application of emission-control equipment.

Rain and cloud water has become more acidic as a result of the burning of fossil fuels that contain sulfur dioxide ($SO_2$) and nitrous oxides ($NO_x$). These pollutants recombine with oxygen molecules in cloud droplets to form sulfuric and nitric acids. These newly formed acids are flushed from the atmosphere through precipitation in the form of rain, snow, sleet, or fog. The highest concentration of acid is found in cloud water that has not yet been diluted. In mountainous regions, such as the Appalachian Mountains, acid fogs have weakened many trees, making them susceptible to drought, disease, and insect infestation.

A primary constituent of contemporary urban smog episodes is ozone ($O_3$), a poisonous substance that irritates the eyes and respiratory system, and retards plant growth. In the lower atmosphere ozone forms in a chemical reaction that involves nitrogen dioxide ($NO_2$) and molecular oxygen ($O_2$) in the presence of ultraviolet light from the sun (thus the term photochemical smog). Ozone concentrations in the lower atmosphere tend to display a daily afternoon maximum and early morning minimum.

Ironically, ozone that occurs naturally in the stratosphere absorbs high energy ultraviolet light in the upper atmosphere, thus blocking these harmful rays from reaching the Earth's surface. A decrease in the concentrations of stratospheric ozone globally (Fig. 9.1) due to a complex chemical interaction involving chlorofluorocarbons (CFC) that depletes stratospheric ozone has raised concerns of increased incidence of skin cancer and crop damage.

The Clean Air Act and its amendments address mainly urban pollution and acid rain and include strategies for controlling high ozone levels in the lower atmosphere. It also offers protection of the stratospheric ozone layer by including provisions for recycling CFCs and calling for a ban on the production of CFC's.

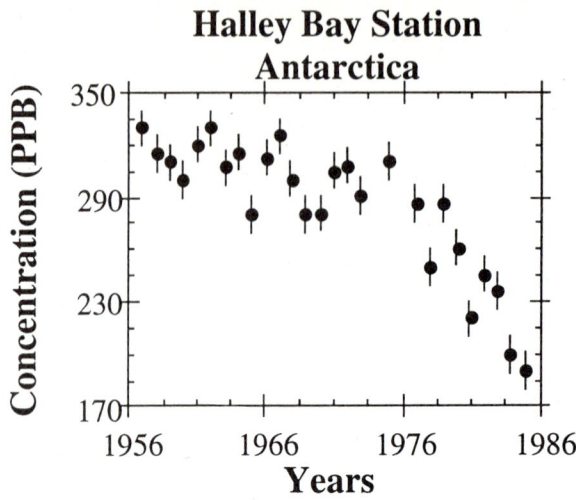

Fig. 9.1 Ozone concentration in parts per billion (PPB) for Halley Bay, Antarctica

*Global Warming*

The global climate of the Earth is largely determined by the balance between available (not counting that reflected back to space by clouds and the surface) incoming short wave (visible) solar radiation, and outgoing long wave (infrared) radiation emitted to space by the Earth and atmosphere. Climate change depends upon changes that affect this radiation balance. Such changes can be separated into several categories:

1. Changes that affect the reflectivity of the Earth-atmosphere system for incoming short wave radiation:
    i) changes in cloud cover (strong radiator in infrared and strong reflector in visible)
    ii) changes in snow cover (strong radiator in infrared and strong reflector in visible)
    iii) changes in the surface characteristics, e.g., deforestation, overgrazing, urbanization
    iv) changes is aerosols, e.g., smoke, ash
2. Changes in the composition of the atmosphere affecting outgoing long wave radiation:
    i) changes in $CO_2$, methane, chlorofluorocarbons
    ii) changes in aerosols, e.g., smoke, ash

3. Astronomical changes affecting incoming solar radiation:
    i) celestial mechanics of Earth's orbit
    ii) solar aging

   The complexity of the climate system is evident in the above outline and our knowlege of the complex interactions between the atmosphere, ocean, cryosphere (ice) and biosphere (plant/animal) systems is incomplete. Our ability to adequately observe the global climate is also insufficient. NASA's Mission to planet Earth and NOAA's Global Change Program include efforts to address these pressing issues.

   Concern for global warming stems from the observed increase in gases such as carbon dioxide and methane, which inhibit outgoing long wave radiation. Records of temperature and carbon dioxide concentration derived from glacier ice cores show a strong relationship between these two. Under conditions of global warming, it is anticipated that the troposphere will warm, the stratosphere will cool, and ocean surface temperatures will rise, leading to more intense storm systems and wider swings in local climate, including enhanced droughts and flood events. Melting of ice sheets and warming of ocean temperatures produce sea level rise, with obvious implication for coastal areas. Observations, though incomplete, seem to support the concern.

   According to a NASA blue print, we can decrease $CO_2$ emissions over the next 30 years by 40% by doing the following five things:

   1. Increase auto fuel efficiency by 30%
   2. Use compact fluorescent bulbs
   3. Produce more efficient motors
   4. Apply tougher standards for refrigerators and dishwashers
   5. Develop alternative renewable energy sources (e.g., wind and solar)

   While human activity (burning of fossil fuels, agricultural activity, etc.) has increased the concentration of carbon dioixde and other radiatively active gases in the atmosphere during this century, it has simultaneously increased the burden of aerosol, decreasing the amount of sunlight reaching the Earth's surface. Recent research has shown that areas where the aerosol input to the atmosphere is greatest have experienced cooler surface temperatures regionally. This effect may partially offset global warming and help explain why observed indications of global warming have been less than those forecast by global climate models.

# Lab 34: Exhausting Problems

## INTRODUCTION

For many high school students, one of the first signs of adulthood is being able to operate an automobile. But along with this wonderful freedom comes much responsibility. In your Driver's Education classes you learned a great deal about traffic safety, but what is the environmental impact of operating an automobile? How much does your automobile contribute to our air pollution problems?

In the introduction to this section a number of serious pollution problems directly related to the exhaust and operation of motor vehicles were mentioned. To summarize here, automobile exhaust contributes to global warming through the emission of carbon dioxide; to acid rain and smog through the emission of nitrous oxides, carbon monoide and unburned hydrocarbons; and to stratospheric ozone depletion through leakage of chlorofluorocarbons from air conditioning units.

## ACTIVITY

OBJECTIVE: In this activity, students will analyze the impact of their motor vehicle habits and the impact of pollution control devices.

PROCEDURE:

1. Fit a clean white sweat sock snugly over the cool tailpipe of your car, using a broad rubber band to keep it in place. Then turn on your ignition and let the car run for three minutes. Turn off the engine and carefully remove the sock. Inspect the sock for particulate matter.

2. Repeat the instructions in 1) with a second clean sock and a car that does not have emission control devices (pre-1975). (If your car has no pollution control devices, choose a second car that does for this part.)

3. Compute your annual fuel consumption by completing the following calculations.
   Number of miles driven per week (avg=200)
   a) _____

Estimated fuel mileage of your car (avg=20 mpg)

b) _____

Divide line a) by line b).
This is your weekly fuel consumption in gallons.

c) _____

Multiply line c) by 52. This is your annual fuel consumption in gallons.

d) _____

We know that gasoline is actually a very complex mixture of chemicals that when burned releases another complex mixture of gases. In one gallon of gasoline, there is the following amount of air pollutants. For each pollutant below, use your annual fuel consumption figure (line d) to calculate your annual release of pollutants from your automobile.

## Data Table

| POLLUTANT | POUNDS PER GAL | X | ANNUAL GALLONS (line d) | = | MY ANNUAL RELEASE (in lbs) |
|---|---|---|---|---|---|
| | | | | | |
| $CO_2$ | 20.000 | X | | = | |
| $NO_2$ | 0.110 | X | | = | |
| CO | 2.300 | X | | = | |
| Hydrocarbons | 0.200 | X | | = | |
| Aldehydes | 0.004 | X | | = | |
| Particulates | 0.012 | X | | = | |
| Organic acids | 0.004 | X | | = | |
| $SO_2$ | 0.009 | X | | = | |
| | | | | | |
| TOTAL | 22.639 | X | _____ | = | _____ |

<u>QUESTIONS:</u>
1. Describe the difference between the appearence of the two socks?  Can you estimate the percentage increase in the particulate pollution?

2. Compare the socks in your class.  Can you make a statement about the appearance of the socks in the class on the basis of the age of the cars, their fuel efficiency, etc.?

3. Based on your calculations above, is your contribution to air pollution significant?

4. There are over 140 million automobiles in the United States.  What would be the total amount of air pollution from automobiles if everyone drove the same as you?

5. Describe four ways in which you could reduce your automobile air pollution.

   a.

   b.

   c.

   d.

# Lab 35: Understanding Acidity

## INTRODUCTION

We have all heard the term "acid rain," but do we really understand what it is and how it affects the environment? In order to understand acid rain it is necessary to understand the pH scale.

The pH scale was developed in 1909 by a Danish biochemist. The pH scale measures the number of hydrogen ions in an aqueous solution and determines the acidity or alkalinity of a solution. Values on this scale range from 0 to 14. A solution with a pH of 7 is neutral; acidic solutions have pH values below 7 and basic solutions have pH values above 7. The pH scale is logarithmic, which means that a change in pH by one number actually represents a change by a tenfold. For example, a solution with a pH of 5 is ten times more acidic than a solution with a pH of 6 and one hundred times more acidic than a solution with a pH of 7.

Very few substances are neutral (have a pH of 7). Distilled water is considered neutral. Background rainwater falling from clean air is not neutral; it has a pH value of 5.6. This is a result of the presence of carbon dioxide in the atmosphere, and the formation of carbonic acid in cloud water. Acid rain is defined as a solution with a pH less than that of background rainwater. Typically, pH values for acid rain range from 4.0 to 4.6. However cloud water with a pH as low as 2.3 has been observed in fogs in the Appalachian Mountains.

## ACTIVITY

OBJECTIVE: The purpose of this activity is to investigate the pH of various acids and bases.

MATERIALS:

| | |
|---|---|
| § aspirin | § milk- fresh |
| § sugar | § milk- sour |
| § cocoa | § vinegar |
| § indigestion tablets | § rain water |
| § salt | § orange/ Lemon/ Apple juice |
| § soap | § soda |

§ tea/ Coffee       § distilled water
§ pH paper          § tweezers
§ small cups        § stirring rod

## PROCEDURE:

1. Add one teaspoon of each solid to one-quarter cup of distilled water and stir until dissolved. For any of the given tablets, add one tablet to one-quarter cup of water.
2. Use tweezers to dip the pH paper into each solution. Do not re-use the pH paper.
3. Determine the pH of each solution based upon the color scale. Record your observations and results in the table provided. Be sure to label each substance as an acid, a base or neutral.

## QUESTIONS:

1. Which solution was the most acidic? Basic?

2. What would happen to the pH value if you dilute the liquid solutions with distilled water?

3. What would have happened to the acidity of the solution if the solids had not been dissolved in water?

4. What would happen to the pH of a solution if it were left uncovered for a weekend?

5. Why are the lowest values of pH observed in cloud water rather than in rainwater? What are the consequences of that observation to alpine forests that exist at an altitude above cloud base?

## Data Table

| Substance | pH Value | Acid or Base | Observations |
|---|---|---|---|
| Aspirin | | | |
| Sugar | | | |
| Cocoa | | | |
| Indigestion Tablets | | | |
| Salt | | | |
| Soap | | | |
| Tea/ Coffee | | | |
| Milk- fresh | | | |
| Milk-Sour | | | |
| Vinegar | | | |
| Rain water | | | |
| Juice | | | |
| Soda | | | |

# Lab 35: Monitoring Acid Rain

## INTRODUCTION

The acidity of rain water depends upon a combination of naturally occurring acids in the atmosphere and acids that are the result of man made pollutants introduced into the atmosphere from a variety of sources. Three types of acid commonly found in rain water include carbonic acid, nitric acid, and sulfuric acid. The concentration of carbonic acid depends upon the presence of carbon dioxide; and gives rain water its natural background acidity (pH = 5.6). Sulfuric and nitric acids are the product of sulfur dioxides and nitric oxides introduced into the atmosphere by industrial power plants and motor vehicles, respectively.

## ACTIVITY

OBJECTIVES: The goals of this activity are to learn how to take pH measurements of liquid precipitation under field conditions, and to gain knowledge of the variation in acid rain episodes, analyze the human impacts on the environment, and become aware that the degree of accuracy of measurements has to be taken into account when results are interpreted.

MATERIALS:
- § pencil, greenhouse type
- § 2 pocket type, student thermometers, single or double scale
- § heavy duty, 1 m$^2$ plastic tarp
- § 500 mL, plastic graduated cylinder
- § plastic rain gauge to measure up to 15 cm. of rain
- § 10+ pH sticks, wide range (permanent type, ColorpHast$^R$)
- § 18+pH sticks, narrow range (permanent type, ColorpHast$^R$)
- § storage bag/vial to store used, dry pH papers, (color is permanent)
- § waterproof plastic case
- § plastic laminated pH Colorchart, wide range
- § plastic laminated pH Colorchart, narrow range
- § map of expedition area

PROCEDURE:

1. Before or at the first hint of rain, spread the 1 m$^2$ plastic tarp on the ground, in an open area, away from trees, and anchor it.

2. Position the rain collecting graduated cylinder so that it catches rain water run-off from the tarp. This will require some ingenuity and you may have to rearrange your anchors. Keep tarp clean.

3. Place the rain gauge in the same general area as the tarp.

4. At intervals throughout the rainstorm, collect water samples and record the sample data.

5. Record the following information on a copy of the Data Sheet that follows:

   A. Record name or names or members in observation group

   B. Date sample taken.

   C. Describe the rain sample site location and vegetation. Provide as much detail as you can.

   D. Record actual location with map coordinates or attach a map to the Data Sheet with the sample site location clearly marked.

   E. Note the duration of rainfall. Time Began, Time Stopped, and Total Time of Precipitation.

   F. Describe the cloud type if you can or describe sky conditions: completely overcast, thunderstorms, passing showers, etc., or use a word description of your own. You can check it against a weather guide when you return home.

*If it rains long enough, attempt to collect at least three samples during the rainstorm's duration. Record the following data for each sample:*

6. Record the following information on a copy of the Data Table:

   G. Time sample taken.

   H. Rain intensity - i.e., downpour, light, medium, heavy, mist, etc.

   I. Wind direction.

   J. Wind speed (best estimate). (Beaufort scale)

   K. Air temperature.

   L. Amount of rainfall accumulated per sample (Rain Gauge).

   M. Determine and record the pH. Use the wide range pH paper which gives the pH to within one pH unit (Record). Obtain a more accurate pH reading by using the appropriate narrow range pH paper. Take three readings and record them on the Data Sheet.

N. Determine and record the overall pH of the entire rainfall by testing the rainwater collected in the rain gauge during the rain. Record.

O. Record total accumulation (Rain Gauge).

P. Dry the used pH papers, label them with the sample number, and place them in the used pH storage vial. Do <u>not</u> discard used pH papers wet or dry on site. Pack them out.

*Check the local pH index, if one is available for the area.*

# Data Sheet

**I. GENERAL DATA**                    **NAME**_____

1.    Date Sample Taken _____

2.    Site Description

      _____

      _____

      _____

      _____

3.    Map Location:

      Coordinates _____
          (Or attach map with location marked)

4.    Rainfall Duration:

      Time Rain Began _____

      Time Rain Stopped _____

      Total Time of Precipitation _____

5.    Cloud Description/Sky Conditions _____

      _____

6.    pH of Total Accumulation _____

7.    Total Rain Accumulation (run-off) _____ cm.

8.    Total Rainfall (measured in rain gauge) _____ cm.

## Data Table

| Sample # | 1 | 2 | 3 | 4 | 5 |
|---|---|---|---|---|---|
| 1. Time Sample Taken | | | | | |
| 2. Rain Intensity | | | | | |
| 3. Wind Direction | | | | | |
| 4. Wind Speed | | | | | |
| 5. Air Temperature | | | | | |
| 6. Sample Accumulation | | | | | |
| 7. pH of Sample | | | | | |
| a. Wide Range | | | | | |
| b. Narrow Range | | | | | |
| Trial 1 | | | | | |
| Trial 2 | | | | | |
| Trial 3 | | | | | |
| 8. Local pH Index (if available) | | | | | |

QUESTIONS:

1. How did the acidity of the samples vary with location and wind direction. Can the observations be explained in the context of the location of pollution sources in the area?

2. How did the acidity of the samples vary with the time the sample was taken; early or late in the day, early or late in the rain event. Can these differences be explained in the context of diurnal variations in pollution or the impact of precipitation on the quality of the air?

3. Investigate the effects of soot/particulates from a gas engine's exhaust, placed on fabrics and moistened with distilled water.

5. How do the leaves of a plant react to the application acids of varying pH? Examine the tensile properties of yarn exposed to acid solutions of varying pH.

# Lab 37: The Atmosphere Effect and Global Warming

## INTRODUCTION

When visible light passes through large windows of a greenhouse, it is largely absorbed after striking objects within. These objects heat up (and in the case of plants result in photosynthesis) and in turn conduct heat into the air of the greenhouse as sensible heat (heat you can sense with a thermometer). The glass or plexiglass windows prevent the building sensible heat inside the greenhouse from escaping and mixing with cooler outside air and the building stays warm even on cold sunny days. Unlike a greenhouse, the atmosphere does not have a barrier that limits mixing of the air, such as the glass in a greenhouse, thus the term "greenhouse warming" when applied to the atmosphere is actually a misnomer.

How does the atmosphere work? Recall that light is emitted by all objects in wavelengths and energies that depend on the temperature of the object. As the Earth's surface is warmed by sunlight, it emits increasing amounts of infrared radiation, radiation with a longer wavelength than sunlight. Several gases in the atmosphere absorb and reradiate this energy back to the surface and are mislabeled "greenhouse gases." Examples include water vapor ($H_2O$), carbon dioxide ($CO_2$), and methane ($CH_4$). Water vapor is the most important greenhouse gas in the Earth's atmosphere. The latter two gases are now increasing daily because of exhaust from cars and burning coal. At present, the atmosphere traps only a small part of the infrared radiation that is emitted into the air at the surface of Earth. As the amounts of carbon dioxide and methane increase, the Earth will become gradually warmer as the atmosphere absorbs and reradiates more and more of the infrared radiation that would otherwise have passed directly out to space.

The temperature here tonight on Earth depends upon several factors -- seasons, latitude, altitude, proximity to oceans, and prevailing weather patterns. But the temperature on some of our closest planet "neighbors" is thought by many to be simply a result of that planet's distance from the sun. As we hope to demonstrate in part two of this exercise, there is another important factor.

## ACTIVITY

OBJECTIVE:  The objective of this activity is to observe and to investigate the processes that occur in a greenhouse and compare them to those that take place in the Earth's atmosphere.  Part two of this activity will compare and contrast the environments of the other terrestrial (Earth-like) planets in our solar system with each other and Earth.

MATERIALS:
For the class:
§ one large bag of potting soil
§ one box of plastic wrap


For each student or group of students:
§ one rubber band
§ two thermometers
§ two large disposable plastic cups
§ hole punch

PROCEDURE
1. Use the hole punch to make a hole big enough for a thermometer to be inserted about an inch from the top of each plastic cup .
2. Fill each cup with dirt  until the soil is about one inch below the hole just made.
3. Insert a thermometer through each hole so that the bulb is about one inch above the dirt and centered in the middle of the cup.  **Caution:  Do not force the thermometer through the hole.  If it will not go, punch a bigger hole.**
4. Turn the thermometer so that it can be read.
5. Cover one cup with plastic wrap, and leave the other cup uncovered.  Secure the plastic wrap on the cup with a rubber band as shown below in Fig. 9.2.
6. On a sunny day, take the two cups outside at the beginning of the class period, and place them where they will not be disturbed.  Stabilize the thermometers so they will not move.

Fig. 9.2

7. Record the initial temperature on each thermometer in the Data Table below.
8. Record temperatures every 5 minutes for 30 minutes.
9. Make a graph for the temperatures in each cup on the same sheet of graph paper. Graph the temperatures on the vertical scale and the time on the horizontal scale. Designate each line as being for the covered cup or the uncovered cup.

**Data Table**

|  | COVERED CUP | UNCOVERED CUP |
|---|---|---|
| 0 min. | ° | ° |
| 5 min. | ° | ° |
| 10 min. | ° | ° |
| 15 min. | ° | ° |
| 20 min. | ° | ° |
| 25 min. | ° | ° |
| 30 min. | ° | ° |

QUESTIONS:

1. Describe what happened to the temperature in each cup.

2. Given that infrared radiation will pass freely through the plastic wrap over the top of the cup, what is making the temperature rise more rapidly in the covered cup?

3. In a glass greenhouse the infrared rays emitted within are trapped to some extent by the glass, not so in a plexiglass greenhouse, yet the latter still stays warm. What does this tell you about the importance of the infrared radiation in keeping a greenhouse warm?

4. In the absence of other factors, what will be the result for global surface air temperatures if greenhouse gases continue to increase?

5. Extra credit: What factors may help offset global warming?

6. Extra credit: Potting soil tends to be moist. How might the results of this experiment been different if dry soil or sand had been used instead? (See the introduction to section 3 for a hint.)

*PART II:* In part two of this activity, we will examine the four terrestrial (Earth-like) planets -- Mercury, Venus, Mars and Earth. Table 9.2 is a chart of data concerning these planets. Please take a moment to review these data then answer the questions that follow.

Table 9.2

| PLANET | DISTANCE FROM SUN (X1,000,000 KM) | LOWEST TEMP (°C) | HIGHEST TEMP (°C) | TEMP RANGE (°C) | ATMOS. |
|--------|-----------------------------------|------------------|-------------------|-----------------|--------|
| Mercury | 58 | -173 | 427 | 600 | None |
| Venus | 107 | 430 | 480 | 50 | $CO_2$ |
| Earth | 149 | -73 | 50 | 123 | $N_2$, $O_2$ |
| Mars | 223 | -125 | -60 | 65 | $CO_2$ |

QUESTIONS:

1. Which planet has the coldest nighttime temperatures? Explain?

2. Which planet has the highest daytime temperature? Explain?

3. Which planet has the largest temperature range? Explain?

4. Which planet has the smallest temperature range? Explain?

5. What two factors do Venus and Mars have in common?

   a.

   b.

6. Why do you think Mercury is so cold at night?

7. From the data above, what do you think the effect of a high $CO_2$ atmosphere is on temperatures observed?

8. What do you think would happen to the Earth's daily temperature cycle if we suddenly lost our atmosphere?

9. What do you think would happen to the Earth's global climate if the percentage of $CO_2$ in the Earth's atmosphere were to double over the next 30 years?

# *Lab 38: Volcanic Eruption*

## INTRODUCTION
Volcanoes are a large source of atmospheric pollution. Mt. Kilauea on the Island of Hawaii expels as much pollution in the form of sulfur gases into the atmosphere annually as the combined output of all coal-burning electric plants in the United States. Scientists are begining to monitor this pollution source globally using satellite data. Catastrophic eruptions such as Mt. St. Helens in Washington state in 1980, and Mt. Pinatubo in the Philippines in 1991 can inject significant ash and sulfur gases into the stratosphere, where resulting fine particles can remain suspended for periods up to several years. These ash clouds block a portion of the incoming solar radiation from reaching the surface, resulting in cooling of the lower atmosphere. In the imediate proximity of a volcanic eruption the ash plume presents a distinct hazard to aircraft by causing jet engines to seize up after they ingest ash. The ash accumulates in a smooth blanket several inches deep down wind of the mountain (Fig. 9.3).

## ACTIVITY

OBJECTIVE: The objective of this experiment is to observe and analyze the dispersion of ash from a model volcano.

MATERIALS:
§ ammonium dichromate; an orange-yellow crystalline compound
§ metal cookie tray or plate
§ match/lighter

PROCEDURE:
1. In a well ventilated area outdoors, make a small pile of ammonium dichromate about 5-10 cm across on the metal cookie tray or plate. If there is no ambient wind consider employing a portable fan for this purpose.

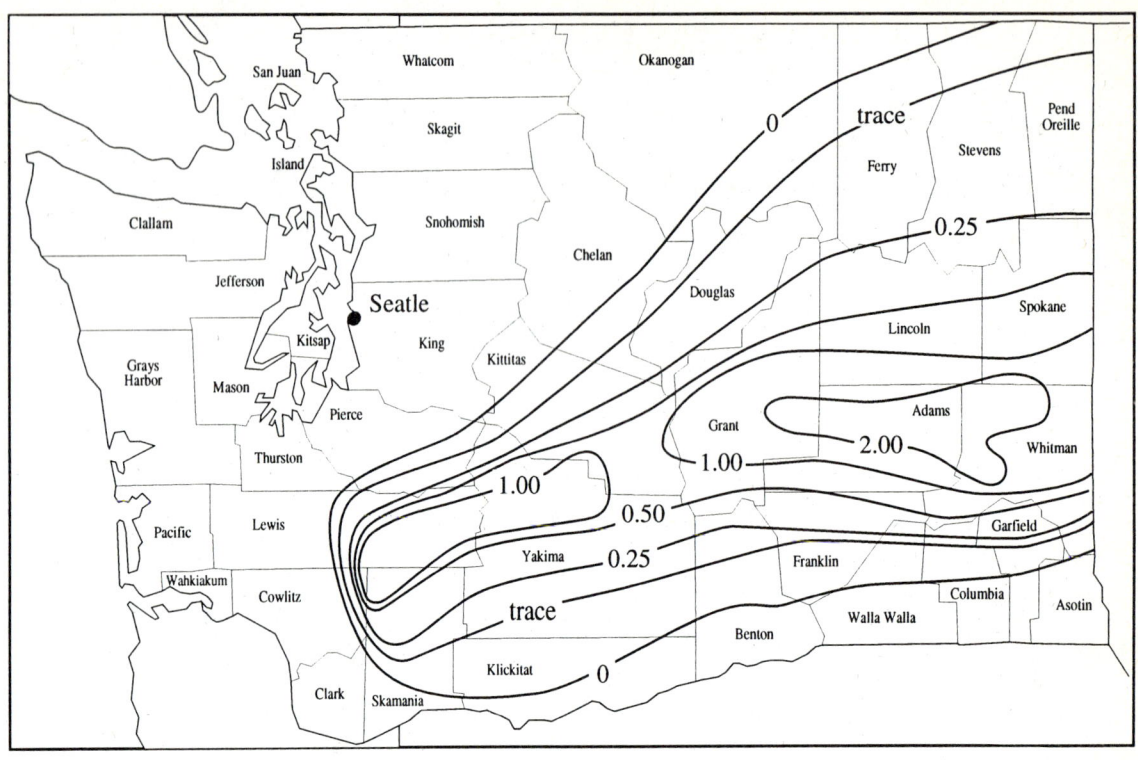

Fig. 9.3  Contours of ash depth following the eruption of Mt. St. Helens on 18 May 1980.  Up to three inches of ash covered large areas of eastern Washington.

2. Light top of the pile with a match or lighter and watch the ammonium dichromate gradually spark and expand.  Although the burn temperature is relatively low, avoid standing too close to the demonstration to prevent inhalation of the ash. The demonstration lasts about 1 minute and provides a nice demonstration of a "strombolian" type volcanic eruption forming a "scoria" (cinder) cone.  **Make careful observations of this process.**

3. Using a pencil or pointed stick make three lines or contours of approximately equal depth in the ash around the volcano.  Choose your contour depths at equal increments (say 1 cm, 2 cm, 3 cm)  to give a reasonable  portrayal of the ash distribution.

QUESTIONS:

1. Carefully describe your observations during the demonstration. What was the ambient wind direction and speed?

2. How was the shape of the contours affected by the ambient wind? How do your contours differ from those in Fig. 9.3? Can you explain the differences?

3. Southerly winds are common over Washington state in the spring. While ash provides good fertilizer for fruit trees in eastern Washington, describe some of the hazards/problems that the ash cloud from St. Helens might have produced under southerly flow?

4. Make a sketch of your final volcano and the distribution of the resulting ash as found from the contours.

**Photograph Credits:**

Cover front and back:  Photograph of the Pacific Ocean, with a winter cyclone visible just west of Kamchatka Peninsula.  A ribbon of convective clouds located just north of  the equator in the lower half of the image is associated with the intertropical convergence zone.  NASA Apollo Saturn, AS13, 11-17 April 1970.

Title page:  Cumulus convection forming over the Island of Oahu in response to solar heating of the ground.  Photograph by the author.

Section 1:  The layered structure of density, cloud water and aerosol content in the lower atmosphere is seen across the sun as it sinks into the Pacific Ocean.  Photograph by the author.

Section 2:  The explosive nature of the eruption of Mt. St. Helens 18 May 1980 was due in part to high pressures created when snowmelt water was heated by lava within the mountain.  Photograph by Austin Post, United States Geological Survey.

Section 3:  Alpine glaciers such as this one near Zermatt, Switzerland provide a record of recent climate swings.  Photograph by the author.

Section 4:  Ice crystals form as frost on a cold window pane in Washington state.  Note how the ice crystals grow at the expense of the much smaller dew drops in the left side of the image.  This is the same process by which rain forms in cold clouds.  Photograph by the author.

Section 5:  Dust and aerosol scatter sunlight over Denver Colorado, resulting in shafts of light called crepuscular rays.  Photograph by the author.

Section 6:   Mountain wave cloud over Boulder, Colorado.  As the name implies, these clouds result from air being lifted over mountains and are stationary.  Photograph by the author.

Section 7:  Heavy snowpack and rimed trees in the Washington Cascades following a series of winter storms.   Photograph by the author.

Section 8:  Hurricane Hugo taken by NOAA GOES East (Geostationary Operational Environmental Satellite) at 5 PM Eastern Daylight Time on 21 September 1989.  The image was created by Steven Chiswell to show the temperature structure associated with Hugo just prior to landfall at Charleston, SC.  The coldest clouds are white and are the tops of deep thunderstorms that make up the eye wall. Warm sea surface temperatures are indicated in the clear eye.  Surface winds at this time were over 125 mph.

Section 9:  A warehouse fire resulting from a natural gas leak produces a thick column of atmospheric pollution over Seattle, WA.  Photograph by the author.